COOL PLANTS FOR COLD CLIMATES

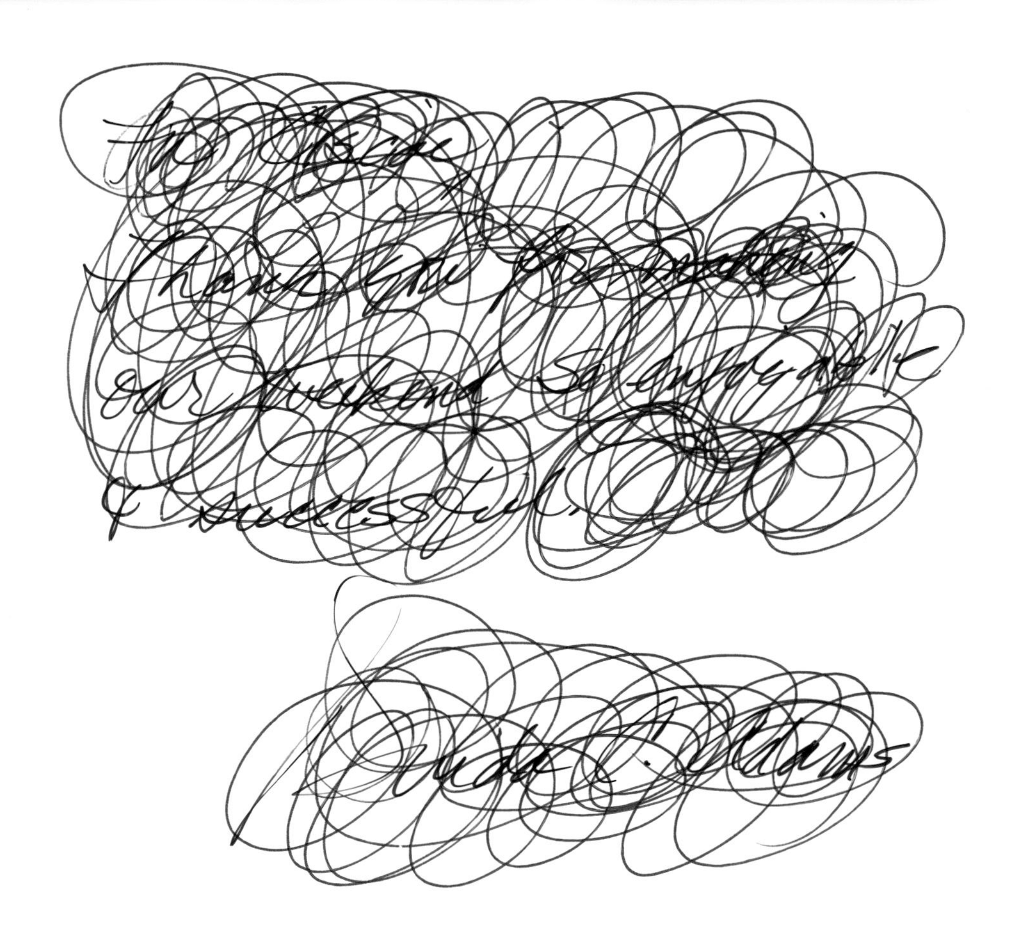

ADVANCE PRAISE FOR *COOL PLANTS FOR COLD CLIMATES*

Brenda Adams, author and designer, has done it again! This time she digs into *Cool Plants for Cold Climates* and gives you the real nitty gritty. As a successful designer, she offers great insights on what will grow in the harshest climates. She discusses everything from bulbs and perennials, to trees and other plants that can grow in this climate. The pictures will inspire you to try some new plants and tackle some new ideas. This book will be an invaluable resource for old and new gardeners because as an experienced gardener and designer, she is one 'cool' writer and fresh new voice!

—**Stephanie Cohen,** lecturer from coast to coast, Fellow of Garden Writers of America, Honor Award Winner from the Perennial Plant Association, author of three books and hundreds of articles, contributing editor and writer for *Fine Gardening* magazine, and certifiable plant geek

I love every page of *Cool Plants for Cold Climates.* Brenda writes with a personal touch that makes us believe we can create gardens as beautiful as those she's designed. She provides not only inspiration, but practical advice for creating a garden with year-round beauty, and even offers pointers on shopping for perennials, trees, and shrubs. This book is a gold mine with nuggets of wisdom on selecting plants for their impact and dependability. An essential companion for every gardener.

—**Julie Riley,** Extension Horticulture Agent, Cooperative Extension Service, University of Alaska Fairbanks

Brenda Adams has created an indispensable resource for cold-climate gardeners, based on the simple premise that they want cool plants, too. *Cool Plants for Cold Climates* is so much more than a plant encyclopedia—it is the sum of the knowledge and wisdom (and wit) of a gardener who loves plants and refuses to be constrained by a short growing season. Adams deftly blends her designer's perspective and gardening expertise into a wonderfully fresh primer for gardeners of all stripes.

—**Richard Hawke,** Plant Evaluation Manager, Chicago Botanic Garden

Brenda Adams presents a garden as a theatrical event, and her scenes encourage gardeners to look beyond flower color to the dizzying array of plant attributes: texture, shape, motion, fragrance, hardiness, and more. Her palette of exceptional plants, practical advice backed by years of experience, and stunning photographs will be inspirational and educational to all northern gardeners.

—**Dr. Patricia S. Holloway,** Professor Emerita, Horticulture, University of Alaska Fairbanks

Finally, a gardening book on plants and landscape design that imparts not just the facts but masterfully conveys the thought process through which expert landscape designers evaluate each plant's usefulness, positive merits, and drawbacks. From the opening sentence of the Introduction—where the author correctly references 'foliage' ahead of 'flowers' in describing any plant's physical makeup—the reader begins to view gardening, plants, and landscape design in an entirely new light.

Each section of the book effortlessly advances the reader's understanding of the myriad roles played by annuals, perennials, trees, and shrubs in the creation of beautiful gardens and overall landscapes. Vital professional secrets are revealed about choosing plants for your garden in such extremely helpful sections as 'Winter Display,' 'Size and Scale,' 'Behavior in the Garden,' 'Choosing a Nursery,' and 'Other Shopping Tips' (it's the first gardening book I've read that tells you exactly why to choose a containerized shrub versus the same shrub sitting next to it).

Combined with Adams' friendly, fluid writing style and gorgeous photography, *Cool Plants for Cold Climates* ushers in an entirely new template for gardening books. Brenda Adams gets it, and after reading this book, you will too.

—**Don Engebretson,** The Renegade Gardener™, award-winning garden writer, speaker and landscape designer in Minnesota

Brenda Adams has crafted a unique and timely book like no other in contemporary garden literature. *Cool Plants for Cold Climates* is an indispensable guide to selecting and growing the finest plants for northern gardens. Lushly illustrated and engagingly written, sophisticated concepts are explained in accessible prose—detailing plant evaluation criteria, cultural requirements, maintenance, seasonality, and aesthetics. Every consideration is expertly explained, from foliage and flower to bark and berry. Rounding out the volume is an encyclopedia of time-tested plants, showcasing the best species and cultivars for ease of culture, durability, and beauty in cold climes. *Cool Plants for Cold Climates* is just the ticket for wintertime garden planning while the snow is flying, a spring foray to the nursery, or any day when you need a dose of encouragement and inspiration.

—**C. Colston Burrell,** author of *Perennial Combinations* and *Hellebores: A Comprehensive Guide*

After Brenda's first book, *There's a Moose In My Garden,* I'm not surprised to find that *Cool Plants for Cold Climates* is absolutely fabulous! In it Adams takes the next step, honing in on the specifics of how to evaluate, grow, and use cold climate varieties for excellent results. Her precise descriptions are easy to follow and understand. With spectacular photos and her usual sense of humor shining through on every page, this is definitely not just another 'how to' book. A definite for every cold climate gardener's library.

—**Rita Jo Shoultz,** Alaska Perfect Peony, www.alaskaperfectpeony.com

Cool Plants for Cold Climates is packed with practical advice as well as plant recommendations to give even the most timid cold climate gardener courage. Adams' passion comes through as she paints vivid word pictures of the plants she features accompanied by excellent photographs. This is a read that's sure to lead to rejuvenation of the gardener's soul on the worst winter's day!

—**Tracy DiSabato-Aust,** bestselling author *The Well-Tended Perennial Garden, The Well-Designed Mixed Garden,* and *50 High-Impact, Low-Care Garden Plants*

COOL PLANTS

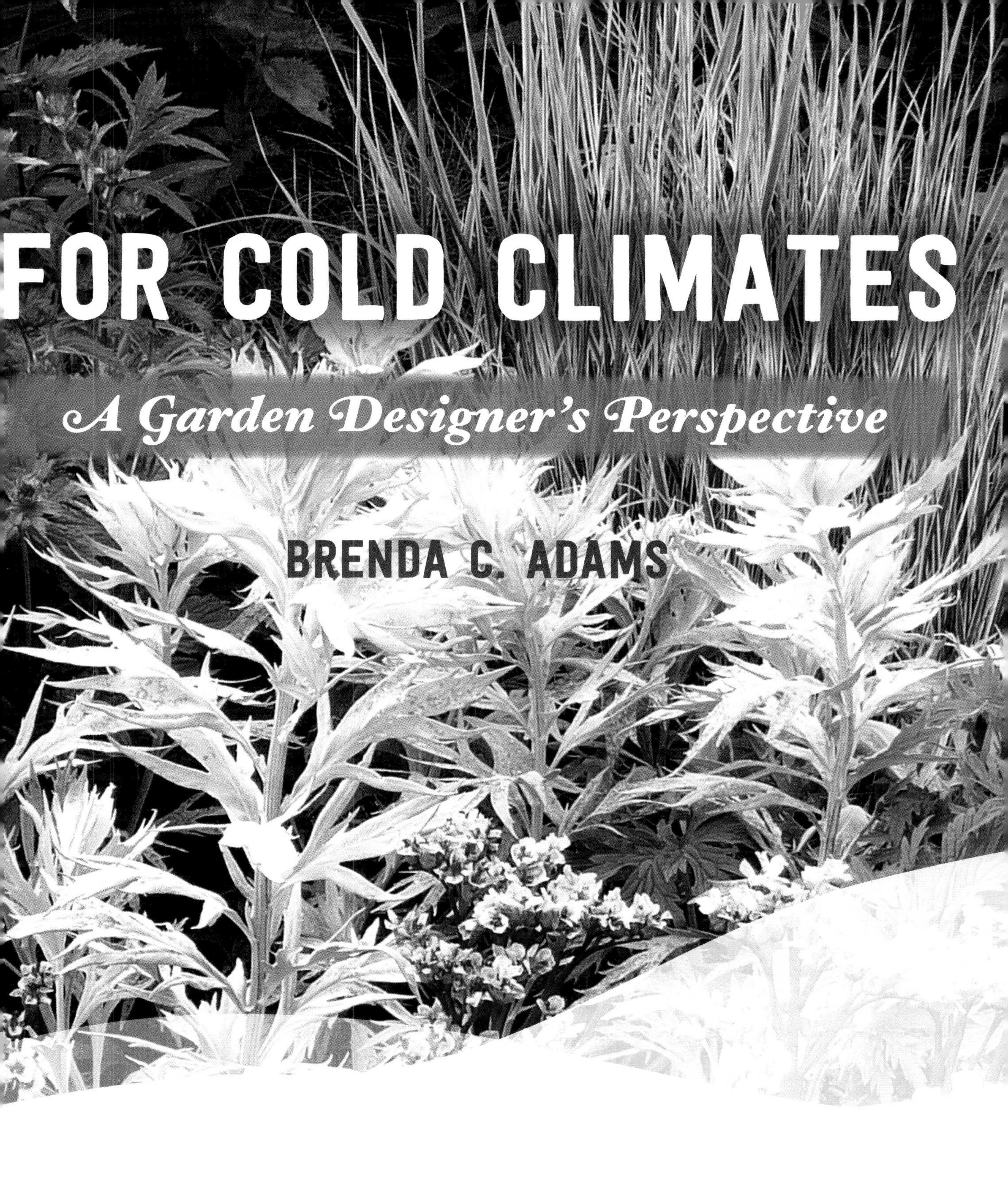

FOR COLD CLIMATES

A Garden Designer's Perspective

BRENDA C. ADAMS

UNIVERSITY OF ALASKA PRESS | FAIRBANKS

Published by
University of Alaska Press
P.O. Box 756240
Fairbanks, AK 99775-6240

Cover design by UA Press.
Interior design by Dixon Jones.
All photos and garden designs by Brenda C. Adams unless otherwise noted.

Library of Congress Cataloging-in-Publication Data

Names: Adams, Brenda C., author.
Title: Cool plants for cold climates / by Brenda C. Adams.
Description: Fairbanks, AK : University of Alaska Press, [2017] | Includes bibliographical references and index.
Identifiers: LCCN 2016056621 (print) | LCCN 2016058296 (ebook) | ISBN 9781602233256 (pbk. : alk. paper) | ISBN 9781602233263 (ebook)
Subjects: LCSH: Cold tolerant
plants. | Plants – Effect of cold on. |
Vegetation and climate.
Classification: LCC QK756 .A27 2017 (print) | LCC QK756 (ebook) | DDC 571.4/642–dc23
LC record available at https://lccn.loc.gov/2016056621

PRINTED IN CANADA

CONTENTS

Acknowledgments *ix*
Foreword *xii*
Preface *xiv*
Introduction *1*

Section I:
Evaluating Plants for Their Garden Impact **6**

Foliage—Shape, Size, Texture, and Color 9
Flowers — Shape, Size, Texture, and Color 13
Bark and Stems 17
Seedpods, Berries, and Other Fruit 21
Architectural Shape and Form 25
Motion 29
Fragrance 31

Section II:
Evaluating Plants for Their Utility and Dependability **34**

Early-Season Value and Fall Beauty 37
Winter Display 41
Bloom Time and Length of Bloom 43
Attractiveness to Pollinators 45
Size and Scale 49
Behavior in the Garden 51
Vigor and Suitability 55
Ease of Maintenance 57
Do You Love It? 59

Section III:
Determining Whether a Plant Will Thrive in Your Garden **60**

Understanding Your Growing Environment 62
Cultural Requirements 65

Section IV:
At the Nursery **70**

Choosing a Nursery—If You Have a Choice 75
Local versus Online 77
Selecting the "Best of the Best" Plants for Your Garden 79
Other Shopping Tips 83
Giving Late Bloomers a Second Look 93

Section V:
Cool Plants for Cold Climates **96**

Annuals 101
Bulbs 107
Grasses 115
Herbaceous Perennials 123
Shrubs 179
Trees 195
Vines 209

Glossary *215*
Materials and Works Consulted *217*
Index *219*

ACKNOWLEDGMENTS

Once again I find myself overwhelmed by the incredible generosity that so many friends, clients, and colleagues have shown to me by contributing their precious time, thoughtful guidance, and invaluable expertise to the creation of this book. Their help and support have made it so much better than it could possibly have been without them.

My dear friend, Roni Overway, reprised her role as editor extraordinaire as she carefully read and reread every draft, gently offering innumerable detailed comments, edits, and suggestions. Her gift of so much time coupled with her superb grasp of the English language can never be repaid. Thank you so much Roni.

A special thank you goes to fellow gardener, best-selling author, and good friend, Tracy DiSabato-Aust, who graciously agreed to preview this book and to write a forward for it. Her insightful and enthusiastic comments perfectly capture its essence and skillfully set the stage for what follows.

Because this is a University of Alaska Press book, it also went through a peer review process. I am so thankful that the university selected such talented and knowledgeable folks as peer reviewers: Verna Pratt, author of several books and arguably the most knowledgeable plant expert in Alaska before her recent passing, Julie Riley, long-time senior Cooperative Extension agent and professor of horticulture, and Rita Jo Shoultz, founder and original owner of the ground-breaking nursery, Fritz Creek Gardens, all reviewed my work. Their suggestions and advice added immensely to the quality of what you now hold in your hands. In addition, Dr. Patricia Holloway, Professor Emerita of Horticulture at the University of Alaska Fairbanks and former director of the Georgeson Botanical Garden, reviewed the manuscript twice for technical accuracy and completeness. Her meticulous review was enormously helpful and is deeply appreciated. Julie Riley and Dr. Holloway also reviewed the plant list for viability in Anchorage and Fairbanks respectively.

What would the world be without gardening friends? I am blessed with a great number of them, many of whom deserve a heartfelt thank you for their assistance on a range of subjects from reviewing the plant list for reliability in various climatic regions to confirming plant identifications and cover design selection, even suggestions for titles. Many thanks to Dan Hinkley, Lucy

Hardiman, Kathy Bingman, Jane Baldwin, Susie Zimmerman, Ed Buyarski, Joan Splinter, Teena Garay, Fran Durner, Carol Swartz, Sharon Froeschle, Gari Sisk, Amy Morton, Jeannette Lawson, Karin Marks, Barbara Kennedy, Lorna Olson, Lee Post, Jenny Stroyeck, Jocelyn Eisenlohr, Cole Burrell, and the late Verna Pratt.

As you will learn from the photo captions, many people welcomed me into their gardens where I was able to photograph excellent examples of "Cool Plants." Some of these folks are clients who invited me to help them create the gardens they envisioned. Their trust, active partnership in the creative process, willingness to let me experiment just a bit, and then to track and photograph the success of exceptional plants in their lovely gardens has been invaluable for this endeavor. They deserve a special thank you. Many of these folks have become friends as well as clients – an unexpected reward. Thanks to Kathy and Mike Pate, Trisha and Ken Pritikin, Gari and Len Sisk, Marguerite and Flip Felton, Denice and Roger Clyne, Leah Evans Cloud, Gail and Bob Ammerman, Lorna and Curt Olson, Jody Murdoch, Cheryl and Cliff Shaeffer, Vicki Rentmeester, Dr. Dots Sherwood (who also cares for our dogs), Elise and Jay Boyer, Cathy and Scott Ulmer, Dorothy and Bill Fry (owners of Bear Creek Winery), the Homer Garden Club, and the ladies of the Auxiliary for the South Peninsula Hospital

The Alaska Botanical Garden, The University of Alaska Anchorage, and the Georgeson Botanical Garden also provided venues where I could photograph many plants, including mature tree specimens. The botanical gardens provide such a vital resource for all of us as do our local nurseries. Thanks to

Alaska Hardy Gardens, The Plant Kingdom Greenhouse and Nursery, Inc., Ann's Greenhouse, and Forget-Me-Not Nursery for letting me photograph their offerings (including less than perfect selections) as a teaching guide for the chapter on plant selection at the nursery. Alaska Hardy, Wagon Wheel, and Forget-Me-Not have also been invaluable as my primary sources for high quality, healthy, and fairly-priced plants for my own garden and for those of my clients. I couldn't do what I do without you folks!

Thank you to the University of Alaska Press team for their professionalism, collegiality, and creativity. Although former acquisition editor, James Englehart, has moved onward in his career, I want to thank him for instantly seeing the value of this book and for pressing the case for its publication. Production editor, Krista West, has been an absolute delight as a partner in bringing this book to fruition. Her creativity, patience, willingness to iterate things until they were perfect, and to counsel with me when my head was bursting, are so appreciated. Several other members of the University of Alaska Press team also made important contributions that need to be recognized: many thanks to Amy Simpson, Laura Walker, and the very talented designer, Dixon Jones.

Finally, but very importantly, is my loving and handsome husband, Bill, who is steadfast in his support and enthusiasm. He reads my manuscripts, keeps the computer humming – no small task – and is always there to offer sage advice and counsel. Did I mention backrubs? Yes, those too. Thanks honey! And thanks to you too, dear reader, for finding my efforts worthy of your exploration.

FOREWORD

I'm beyond excited to see this exceptional, one-of-a-kind, and, yes, very cool book by award-winning garden designer Brenda Adams! Many years ago, I was first asked to speak in Alaska for a master gardener event in *March* in *Fairbanks.* As a winter lover I was thrilled at the prospect of cross-country skiing, watching part of the North American Championship Sled Dog Race, and doing some mushing of my own. My husband brought me back to reality when he asked, "What are you going to talk about? The two plants they can grow in that climate?" Right! Then, when doing research, I found there was really no reference available at that time for extremely cold climates.

As a gardener, designer, and author for mainly the Midwest and eastern states, this was going to be the start of my education on gardening in Alaska and far northern climates. The skiing (although it was –23°F) and lectures were a great success and I have happily returned to Alaska to speak in Anchorage, Kenai, and Homer.

These return visits have, thankfully, been guided by Brenda's very knowledgeable and talented hand and her first book, *There's a Moose in My Garden.* Visits to her spectacular gardens and those of her many clients have taught me about the simply stunning and hardy plants that can be grown in such a cold climate and how incredibly massive they can become in the long days during such a short growing season. This newest book, *Cool Plants for Cold Climates,* draws on her twenty plus years of experience in working with these plants and beautifully illustrates for all of us the wide range of amazing selections available for cold weather gardeners.

Brenda's passion comes through as she paints vivid word pictures of the plants she features accompanied by excellent photographs. This is a read that's sure to lead to rejuvenation of the gardener's soul on the worst winter's day! She provides information on outstanding annuals, bulbs, herbaceous perennials, ornamental grasses, vines, shrubs and trees enabling the reader to create a hardy, artistic, and gratifying mixed garden in the harshest of environments.

Just like when she and I talk and she explains in her engaging manner some of the plants she has used in her gardens, her writing is also fun and engaging. Brenda is a great authority on "all things" plant, design, and gardening, yet her writing isn't overly technical but easy and joyful to read. She covers a wide variety of important considerations including the showcasing of seedpods, berries and other fruit, architectural shape and form, bark and stems, flowers, foliage, fragrance, utility and dependability, size and scale, early and late season interest, and, of course, the winter display.

Cool Plants for Cold Climates is packed with practical advice as well as plant recommendations to give even the most timid cold climate gardener courage. You will find information on what traits make a plant exceptional. Attributes such as garden impact, ease of care, artistic aspects, hardiness, attractiveness to pollinators, and more are meticulously addressed.

Because gardening in the north, with all of its challenges, isn't for sissies, Brenda will take your hand and helpfully explain how to determine whether or not a plant will thrive in your environment. However, while it's great to learn about all these plants, it's no use if you can't procure them. Brenda comes to the rescue again with her advice on how to select the best of the best nurseries and other shopping tips; she understands that shopping for plants is really our favorite thing to do as gardeners! And once you get your new treasures home she offers insight on how to transition your new plants from the nursery to your cold climate garden.

Bravo Brenda for *Cool Plants for Cold Climates* and for teaching us all, including my husband, that indeed you can grow more than just "two plants" in cold climate gardens! In fact you can grow numerous gorgeous, hardy, and rewarding plants. I can't wait to return to Alaska to speak, armed with this very cool, outstanding, and useful reference!

—Tracy DiSabato-Aust
Bestselling author of *The Well-Tended Perennial Garden, The Well-Designed Mixed Garden,* and *50 High-Impact, Low-Care Garden Plants*

PREFACE

Since the release of my first book, *There's a Moose in My Garden: Designing Gardens in Alaska and the Far North,* I've been invited to speak to a wide range of audiences from small garden clubs to Master Gardener groups and classes, from Botanical Garden conferences to the country's second-largest garden-related venue, the Northwest Flower and Garden Show. This has been an absolute joy for me because I love teaching and sharing what I've learned over the years, whether I came by the knowledge through study, experience and experimentation, or the generous sharing of other gardeners.

Invariably, even though the topics of my presentations have most often focused on some aspect of garden design, the question I hear most frequently, as I project photo after photo of beautiful gardens onto the screen is, "What *is* that plant?" Someone's interest has been sparked by the frothy pink blooms in the foreground, the tall, willowy grass on the left, a deep burgundy shrub in the background, or simply a clutch of vibrant orange berries. Perhaps it's the otherworldly quality of an unusually shaped glacier-blue flower, a brightly-colored ground cover, or the big, bold foliage of a focal point. Whatever attribute stirred the interest of my listener resulting in a raised hand, ultimately it was a specific plant that generated the excitement. As it turns out, not all gardeners are concerned about whether their gardens are well designed, but all the gardeners I've met want their gardens to be populated with plants they find attractive and appealing. They also want plants that will thrive in their unique location.

While there are multitudes of books about plants that do well in warmer climates, there is little available for truly cold-climate gardeners. For these intrepid souls, finding an exciting plant that is new to them and hardy in Zone 2, 3, 4, or 5 and then learning enough about it to determine what it will add to their gardens, can be a real challenge. If you are one of these courageous gardeners, I dedicate this book to you. It is a book about great *cold*-climate plants, how to evaluate them, how to determine whether they will do well in your garden, and where and how to buy the best of the best, all while supplying you with in-depth descriptions of a broad range of exciting selections.

The descriptions are enhanced by more than three hundred photographs. As noted in the Acknowledgments, many of the photographs in this book were taken in the gardens of friends and clients as well as several public gardens. In the captions I endeavor to accurately identify the garden where I took the photographs. Images with no attribution were taken in my garden in Homer, Alaska, which serves not only as a place of pleasure for my family, but also as my laboratory. I hope you will find the photos helpful in visualizing how these plants might fit into your garden.

Northern gardens need not be uninteresting or fragile: cool—*really* cool—hardy plants are out there. My goal is to provide you and other cold-climate gardeners with enough information about a wide variety of exceptional plants so that you will be able to create a gorgeous, gratifying, and dependably hardy garden, a garden filled with selections you adore. I hope you will find this information both refreshing and valuable.

INTRODUCTION

Not all plants are created equal. Some have the structure to be stunning focal points, while others sport finely textured foliage that provides a subtle backdrop for the delicate flowers of a nearby partner. Each of these roles can be important to the beauty of your garden, but selecting the best candidates is vital. If you think of your garden as a well-crafted play, you become the director who selects the perfect actor for every part, actors who are compatible and have wonderful chemistry. Just as a director would choose a different cast for a drama than for a comedy, your choices should also be well suited to the style of garden you plan to create. For example, you might fill a formal entry garden with quiet, refined foliage plants, but choose colorful, billowy, and exuberant plants for a sunny cottage garden.

Regardless of its style, your garden will likely be composed of a few stars, some key supporting cast members, and many bit players, all of whom come and go throughout the production. Each will sport a unique costume and play a vital role as it appears. Careful placement of each plant, with an eye toward how it will develop over time, will make the entire production evolve beautifully and lead to a long-running success.

In some plays the set is simple; in others it is lavishly detailed. Either will be integral to the way the audience experiences the show. So it is with gardens: many are surrounded by plain fences or dense stands of trees, some are in the foreground of spectacular vistas that lie beyond their borders, and still others are nestled near colorful meadows or abut bustling cityscapes. How you choose to visually incorporate the adjoining scenery into your garden will affect the way folks visiting your garden, including you, will experience it.

Lighting brings out the best in players on the stage and in the garden alike and can add an element of mystery, vitality, even surprise. The low angle of sunlight in northern gardens offers those who garden at high latitudes the special effects of prolonged backlighting and soft ambient light that infuses pastels with intensity and more saturated colors with exceptional liveliness.

Music, too, has a role in theater. The music of your garden might be the melody of moving water or the whisper of swaying grasses, the rustle of crisp foliage or the rattle of seed heads, the songs of visiting birds or the buzz

▲ The iris is clearly the star of this combination, while the perennial bachelor's button (*Centaurea montana*) enhances the star's role by repeating its color and offering contrast with an understated, yet interesting-looking blossom. Molly Stonorov's garden and design.

◄ The design goal of the Serenity Garden at South Peninsula Hospital was to create a calm and restorative environment. To achieve this, we selected plants with restful colors that melded smoothly together just as good supporting actors do.

of busy bees, or even the laughter of children. Each adds a gentle note of beauty to your creation.

How, then, do you evaluate a plant so you can select the best players for your production? What characteristics infuse a plant with star quality or make it a good, solid supporting member of the cast? Which plants are unequivocally the best ones, able to both enhance your garden and thrive in your location? What should you look for in any candidate? In other words, what makes a plant a great choice? These are all important questions. Let's take a look at the answers.

A great plant brings many outstanding qualities to your garden. A good portion of these have to do with the artistic impact and beauty of a plant; others relate more to its exceptional utility and potential for success in the garden. Among the artistic or visual qualities you might consider are these: bold or unusual foliage; striking flowers, buds, or seedpods; colorful or textured stems or bark; appealing fruit; compelling architectural shape; and eye-catching motion. Those that relate to utility and dependability include long bloom time, multiple seasons of interest, early spring or late-season value, attractiveness to all sorts of pollinators, ease of care and good behavior, as well as hardiness and vigor. In a category all its own, but of particular importance to a garden, is an evocative fragrance.

Keeping these qualities in mind, how can you best identify an exciting new plant and then understand enough about it to determine what it will add to your landscape *before* you put it into the ground? Most descriptions available to you will detail a plant's potential height and width and provide a bit about its horticultural needs and its bloom color, and perhaps its bloom time. While useful and necessary, these data are insufficient for you to determine if this possible choice is truly worth your money and effort.

Unfortunately, there has been a lack of sound, qualitative, and descriptive information available for cold-climate gardeners. When I moved to Alaska twenty-five years ago from very warm climates (Southern California and Arizona), my aim was to steep myself in knowledge about what selections would be successful in this dramatically different Zone 3 environment. After searching all the bookstores within 230 miles, I found only one book, Lenore Hedla's *The Alaska Gardener's Handbook,* that addressed the subject at all.

Although what I learned from Lenore was very helpful, it was less detailed than what I was seeking. Friends and neighbors offered suggestions and shared plants, but for the most part, I was left with a trial-and-error process, and, thankfully, sometimes trial and success. I was blessed to live near a creative and ground-breaking nursery, Fritz Creek Gardens, which was focused on expanding the local availability of plants that would thrive in Zones 2–5. I tried just about everything they offered. Some varieties did well; some

didn't. Over time, through seemingly unending experimentation and after careful observation, I developed a list of great plants that were robustly hardy. To create the list, I had to learn much more about each plant than simply whether it would survive a cold, challenging growing environment. I also had to study its character and behavior, and take into account what it added to a garden and to my pleasure.

In the first two sections of this book, we will explore the attributes that I've determined make a plant exceptional, why each is important, and what it will add to your garden's overall beauty and enduring success. As mentioned earlier, some of these traits are purely artistic, while others are more utilitarian. Since I think each is important, I've described them at some length. All are qualities that I considered in determining which plants earned the moniker "cool." Because you will likely want to continue your quest for new and different plants, understanding how to make these judgments yourself will be endlessly helpful.

▲ Lighting is a key element in every garden. In northern gardens, the low angle of light makes pastels more intense and vibrant. As the directors of our gardens, we can use the quality of our light to create special effects.

▲ Not all cool plants are stars. Many are supporting cast members, like this garden pink (*Dianthus gratianopolitanus* 'Firewitch'), which is covered in small, intensely fragrant, fuchsia flowers for weeks on end. It has attractive blue-green foliage and is very easy to maintain. Gari and Len Sisk's garden.

In the third section, you will learn how to evaluate your garden, its soil, and other environmental factors with the goal of learning enough about them to equip you to select plants that are well suited to your environment. This is the fundamental key to having a healthy, thriving garden, thus making the care of it enormously easier.

Now comes the fun part—shopping for your plants. We've all bought plants simply because we found ourselves having a "love at first sight" moment with them—especially when they are in bloom—rather than making a carefully considered selection. That's just fine; gardening doesn't always need to be a serious endeavor. On the other hand, and putting flights of fancy aside, have you ever found yourself in a situation where you didn't know which was the best choice among the available options of a particular kind of plant? I know I have. It took me a long time to learn how to critically evaluate and confidently make the choice between one specimen and the next. That's why I've included Section IV, "At the Nursery." It will guide you through this process, giving you buying tips that will help you know what to look for whether you are selecting an inexpensive annual or making a major investment in a large tree.

Finally, in Section V, "Cool Plants for Cold Climates," we'll conclude with detailed portraits of the best of these hardy plants. I've attempted to shine a spotlight on the outstanding features of each choice, highlighting its artistic value for a garden composition. These descriptions and the accompanying photos are intended to help you better visualize the plant and understand the role it can play in your garden. Because they are of utmost importance, I've also included details about each entry's growth habits and horticultural needs. As you read, you'll learn what makes each a cool plant—a plant that will offer far more than others can to help you create a splendid garden in cold gardening country.

That being said, this book is not intended to be encyclopedic. Instead, it presents many of the genuinely wonderful plants that thrive in most areas of Alaska and other particularly cold regions of the world, supplying zone-based guidance as to whether they have been successfully grown in areas like yours. These are plants that, in addition to their beauty and hardiness, bring many other attributes to inform your creative process.

◄ A few diminutive plants have such quiet or unusual beauty that they demand our full attention. A good example is the exquisite spotted lady's slipper *(Cypripedium guttatum)* in the late Verna Pratt's garden.

SECTION I

Evaluating Plants for Their Garden Impact

As you begin "auditioning" new plants for your garden, understanding how each will enhance your creation is crucial. What is a particular plant's artistic value? Will it be the star, or will it be a supporting cast member? Will it blend well with the plants you already have? Is its growth habit compatible with them? Most importantly, do *you* find it charming and delightful? There's absolutely no reason to add or keep a plant that you find ho-hum or that you've decided you simply don't like.

Because I'm a garden designer, I always consider how the attributes of one selection can complement the features of another if they're placed together in a garden combination. To do this, I need to know each plant intimately. I must understand much more about it than simply its potential size and flower color. Using this approach takes some practice, but when you start thinking in terms of artistic impact, you'll begin to see what makes a particular plant exceptional. You'll look beyond its flowers to see all the other elements that provide color and character. This knowledge can lead to your next step: determining how to best use this new plant to maximize its impact in your garden.

Let's examine the concept and characteristics of artistic impact more closely.

◀ Irresistibly soft lamb's ear (*Stachys byzantina* 'Helen von Stein') with bugleweed (*Ajuga reptans* 'Burgundy Glow').

More about Garden Design

There's a Moose in My Garden, **by Brenda C. Adams (University of Alaska Press)**

The Well-Designed Mixed Border, **by Tracy DiSabato-Aust (Timber Press)**

Perennial Combinations, **by C. Colston Burrell (Rodale Press)**

The Perennial Gardener's Design Primer, **by Stephanie Cohen and Nancy J. Ondra (Storey Publishing)**

FOLIAGE
SHAPE, SIZE, TEXTURE, AND COLOR

Foliage is one of the most powerful features of a plant. Not only is its presence felt throughout the growing season, but it is available in a broad range of shapes, sizes, textures, and colors. If you make use of these qualities, you can transform your artistic endeavors from simply pleasant to absolutely outstanding.

Great plants often have leaves with distinctive texture. Some are so soft and fuzzy that you can't refrain from touching them. I have a clump of lambs' ears (*Stachys byzantina* 'Helen von Stein') in my garden along the edge of a pathway. In the twenty plus years it has been there, not a single child under the age of twelve has ever walked by it without bending down to touch the enticing velvety leaves. At the opposite end of the spectrum are a few plants that have foliage that is clearly prickly in the extreme, warning you to keep your distance while, at the same time, creating an interesting textural contrast. Thick, meaty foliage characterizes a number of plants; delicate and wispy is the hallmark of others. Big and bold, midsized but deeply cut, or fine and colorful—there are a multitude of ways in which a plant's foliage can set it apart from all the rest, and, in combination, add impact and beauty, the kind of beauty that can't be ignored.

In addition to a variety of shades of green, foliage is available in a phenomenal number of diverse and novel colors as well as in many variegated forms. These options can enhance, repeat, and emphasize your color theme, and, at the same time, they possess the ability to add a spark to your garden during periods when flowers are sparse or even absent.

◄ Colorful and succulent stonecrop (*Sedum* 'Vera Jamison').

▲ A variety of shapes, sizes, and patterns adds excitement to a green-and-white combination. Foreground from the left, spotted deadnettle (*Lamium maculatum* 'White Nancy'), lungwort (*Pulmonaria* 'Trevi Fountain'), and hosta (*Hosta* 'Patriot') repeat the white flower color.

▶ Roger's flower (*Rodgersia podophylla* 'Rotlaub') and cushion spurge (*Euphorbia polychroma* 'Bonfire') make an attractive foliage combination, juxtaposing different foliage sizes and shapes while repeating a color theme. Gari and Len Sisk's garden.

FLOWERS

SHAPE, SIZE, TEXTURE, AND COLOR

lowers are often what initially attract us to a particular plant, especially when it's blooming at the nursery or in a friend's garden. Of the many attributes of each blossom, color is probably the first thing most of us notice. Often a flower will have more than one color on its petals while its sepals or stamens may offer additional hues. The shape, size, and texture of flowers play a vital role in distinguishing one plant from another. Although large blossoms quickly command our attention, sometimes it's a plant with small dainty ones that catches our eye, enveloped as it might be in a coverlet of tiny delicate flowers. I particularly enjoy those that face confidently upward with radiant, happy-looking faces rather than the shy-looking, demure ones with downward-glancing, pendant

◀ Massed plantings make the impact of flower color even more striking. Entry garden at Stream Hill Park.

◀ Big, bold, and bright, two-toned *Lilium* 'Royal Sunset', an LA Hybrid, holds center stage while in bloom. Homer Garden Club's Baycrest garden.

heads—but you'll find a number of each to love among the great plants described here.

Like foliage, blossoms have a broad range of textures. They can be soft and delicate, spiky and sharp, meaty and substantial—even waxy and smooth. In addition, there is an endless medley of shapes and configurations, from the bold, dramatic trumpets of Asiatic lilies to clusters of hood-shaped blossoms arrayed on the upper stems of monkshood or the sweet pendant bells of campanula. The trumpets of daffodils, the long spurs of columbines, the elegant spikes of salvia and veronica—even the composite flat ray of a daisy, so attractive to butterflies—all offer gardeners delightful choices. Some of my favorites are the simple, yet perfectly round flower heads of globe thistle (*Echinops*) and those of many of the ornamental alliums.

It's exciting to see the roles that different flower shapes play in attracting pollinators. Butterflies are attracted to daisy-like flowers partly because they provide a nice place to land. Although bees also are attracted to ray-shaped flowers, they will downright swarm the small flowers of catmint, allium, sage, alyssum, bee balm, and veronica, to say nothing of lamb's ears. You might even find them resting in the large tubular flowers of foxgloves. Alluring and always desirable, hummingbirds are especially attracted to red flowers. Their long bills make them uniquely adept at probing funnel-shaped flowers that have deeply hidden pockets of nectar. The low-pitched hum generated by their intensely beating wings is an especially welcome sound of summer.

▶ Backlit spiky speedwell (*Veronica spicata* 'Red Fox').

▲ Flowers totally cover the foliage of two moss phlox cultivars (*Phlox subulata* 'Candy Stripes' and *Phlox subulata* 'Emerald Pink') in Gari and Len Sisk's rock garden.

▶ Delicate-looking flowers atop wispy silver foliage make snow-in-summer (*Cerastium tomentosum*) an engaging ground cover.

BARK AND STEMS

hen searching for plants that merit the adjective "cool," be sure to consider the bark and stems of your possible choices. These can be a subtle but notable source of color as well as texture. Two outstanding examples are the pure white exfoliating bark of a mature white birch (*Betula papyrifera*) or the shiny cinnamon-bronze of Amur chokecherry (*Prunus maackii*) with its raised horizontal markings. The trunks and branches of both of these trees provide color and texture throughout the year, making them desirable stars in the garden during winter and summer as well as the seasons in between.

◂ The bark of some quaking aspen (*Populus tremuloides*) is green, while other specimens are gray, tan, or even nearly white.

Several hardy shrubs also have colorful, woody stems. Red- and yellow-twigged dogwoods stand out brilliantly against winter snow. Dwarf blue arctic willow (*Salix purpurea* 'Nana') sports dark purple stems all year and

◂ Red-twigged dogwood (*Cornus alba*) in Lorna and Curt Olson's garden.

is adorned with unusual dusty gray-green foliage throughout the summer, providing its own stunning combination.

Many herbaceous perennials (those perennials that die back to the ground in winter) also display vibrant colors on their stems. The rocket ligularia (*Ligularia stenocephala* 'The Rocket') is an excellent example with its tall, deep chocolate stems, as are several bronze and orange varieties of cushion spurge (*Euphorbia polychroma*) with their persimmon or tangerine stems. The scarlet-stemmed meadowsweet (*Filipendula purpurea* 'Elegans') is another. So it's not just woody plants that have colorful stems; nature offers us color in a wonderful assortment of options—sometimes subtle, sometimes striking, sometimes absolutely riotous.

▲ Some vegetables display colorful stems, as does this chard from the 'Bright Lights' collection.

◀ A pretty mullein (*Verbascum*) with dark chocolate stems in Shirley and Harry Forquer's garden. Shirley's design.

SEEDPODS, BERRIES, AND OTHER FRUIT

As flowers fade, seeds are formed. **As flowers fade, seeds are formed.** Sometimes these can be as engaging as the flowers themselves. Good examples are the fuzzy white seed heads of pasque flowers (*Pulsatilla*), clematis vines, and clematis shrubs. Many plants produce unique pods, button-like orbs, or large sprays of tiny seeds that can persist into winter, extending the period of fun goings-on in your garden and thus increasing the enjoyment you gain from it.

◄ Dill has attractive seed heads that can be harvested for cooking.

Clusters of bright, crisp berries decorate a variety of plants in the late summer and into fall. These are not only a great source of color but also an important food source for our feathered garden visitors. The perfectly round shape of berries and other fruits is always a welcome addition to any planting, juxtaposed, as they might be, against the spiky or frilly structure of their neighbors. Although not as colorful as many berries, the cones of conifers also add charm to the garden and persist into winter, providing seeds for wildlife.

◄ Water droplets sparkle on the airy seeds that decorate ornamental grasses toward the end of the season and into winter.

▲ The understated cones of conifers add a lovely dimension to winter gardens.

◀ In late summer and fall, beautiful and edible berries adorn American mountain ash (*Sorbus Americana*).

ARCHITECTURAL SHAPE AND FORM

By the "architecture" of a plant, I mean the plant's three-dimensional form. The arrangement of a plant's stems, branches, leaves, and flowers determines its architecture. Some plants have a loose, open architecture that imbues them with a light, airy feel. Ornamental grasses often fall into this category. Conifers, on the other hand, are generally densely branched, creating a solid, weighty impression.

▲ Cushion spurge (*Euphorbia polychroma*) forms a perfect mound.

◄ Dense and somewhat pyramidal in shape, a blue spruce (*Picea pungens* 'Fat Albert') anchors and contrasts with a variety of more loosely structured, mounded plants, including dwarf meadowsweet (*Filipendula* 'Kahome'), a few spiky delphiniums and veronicas, and a colorful ground-covering bugleweed (*Ajuga reptans* 'Burgundy Glow'). Pritikin Family's garden.

▲ Great yellow gentian (*Gentiana lutea*), with its bold, pleated foliage, pyramidal shape, and unusual yellow flowers, can serve as either a garden star or a strong supporting cast member.

Between these two extremes are many interesting configurations, some of which can impart a very strong presence that will hold its own as one of the stars or focal points in the garden. For example, the bold-leaved ornamental rhubarb (*Rheum palmatum* 'Atrosanguineum') is huge and colorful and produces flowering spikes that reach eight feet in height. Or consider the wide selection of columnar, cool-season grasses such as the cultivars of feather reed grass (*Calamagrostis* x *acutiflora*), which are quite eye-catching individually as well as en masse. Additionally, the spiky-looking cardoon or artichoke thistle (*Cynara cardunculus*) easily takes center stage in an herbaceous perennial or herb garden.

To balance the "look at me" quality of very bold plants, there are many others that have a more demure appearance. Some are naturally round, upright, vase-shaped, or prostrate (ground-hugging), while others are triangular, spiky, frilly, mounded, or columnar. This shape or silhouette is part of the artistic value of the plant too, and can be used to add drama, rhythm, or subtle elegance. Orchestrating the interplay of these different shapes is an important step toward making your compositions more interesting and more artistic. A round plant placed slightly forward of a triangular one, for instance, and fronted by a nice prostrate ground cover, is the beginning of a satisfying scene. Add a spiky silhouette and something frilly, and you've created an interesting combination regardless of the other qualities of the plants. Repeating a form or combination periodically creates rhythm and injects cohesion into your landscape.

▲ Vertical, upright Siberian iris (*Iris sibirica* 'Caesar's Brother').

▶ Although we often think of variety as the "spice of life," this collection of mounded, brightly colored sedums uses repetition in an incredibly effective way. Norma Leland's garden and design.

MOTION

When the wind is howling, there's a lot of motion in our gardens—most of it unwanted. On the other hand, the eye-catching nature of subtle movement is endlessly fascinating. It is why moving water can captivate us for hours and why ever-popular wind sculptures are completely mesmerizing.

Grasses might be the first group that comes to mind when we think about undulating plants, but there are many others. Trees that have a weeping form, such as the stately weeping willows along the banks of the eastern rivers of my childhood, sway in just a whisper of a breeze—a real treat. So do the branches of the much hardier weeping white birch (*Betula pendula*). Perennials that have tall, flexible stems, such as Siberian iris (*Iris sibirica*), queen of the prairie (*Filipendula rubra* 'Venusta'), and Culver's root (*Veronicastrum virginicum*), can also contribute motion to your landscape.

Other sources of activity in our gardens are the flitting of birds and the industrious nectar gathering of our many pollinators. Selecting plants that will attract these welcome visitors is another way to increase animation and enhance the pleasure your garden offers.

◀ A mix of grasses bends gracefully in the seaside breeze in Gari and Len Sisk's garden.

▶ Even in winter, *Calamagrostis* x *acutiflora* 'Overdam' rustles in a gentle wind. The South Peninsula Hospital Auxiliary's Serenity Garden.

▶▶ The appropriately named quaking aspen (*Populus tremuloides*) has wonderful foliage that flutters enchantingly.

FRAGRANCE

Have you ever been transported to another place and time by a tiny whiff of a familiar aroma? Fresh corn, an unfortunately rare commodity in Homer, Alaska, does it for me. Inhaling the distinctive fragrance of the silk of a fresh ear of corn transports me to my early childhood and visits to my Aunt Hazel's farm. When corn was on the menu, a big pot of water was already on the stove and boiling as my cousins and I were dispatched to the cornfield to pick the most perfect ears. We dashed back to the house with them at breakneck speed. I can remember burying my face in the treasure in my arms as I ran, anticipating the tasty treat to come. It was heaven! Now, decades later, when I am exposed to that fragrance, I am transported back to the farm, running just as fast as I can in the warm Pennsylvania sunshine.

Our sense of smell has the longest "memory" of any of our senses. For many years scientists have wondered about why and how this works. So have I! Recent research has determined that our olfactory response has a more direct link to our long-term memory than do our other senses. Apparently, the earliest event in which we experience a special smell is laid down in our long-term memory along with its associated fragrance. As a gardener, you may choose a plant for your garden because you associate its perfume with a special moment in your life, perhaps with your grandmother or another cherished friend or relative. Wouldn't it be a wonderful legacy to create the same unique bond with your children or grandchildren? If you believe so, as I do, be sure to explore the world of fragrant plants and include them in your garden.

There are many other reasons to search out great plants that have wonderfully evocative fragrances, not least of which is the pleasure they will bring *you* in that moment when you take the time to sit among them or as you pass them by while strolling in your garden, delighting in what you've created.

◄ A particularly pretty lilac (*Syringa vulgaris* 'Sensation') has white-edged flower petals and a light fragrance. The Pritikin Family's garden.

▲ Incredibly fragrant *Narcissus* 'Manly' is but one of many spring bulbs that can also add their perfume to the garden.

▲ For over-the-top perfume, consider growing 'Blizzard' mock orange (*Philadelphus lewisii* 'Blizzard').

▶ Many pinks are sweet-smelling, including this two-toned one, *Dianthus* 'Coconut Surprise', which has a clovelike scent. The Pritikin Family's garden.

◀ *Viola* 'Etain' has a mild spicy aroma. To enjoy its subtle fragrance plant it along a path, beside a garden bench, or in a nearby container.

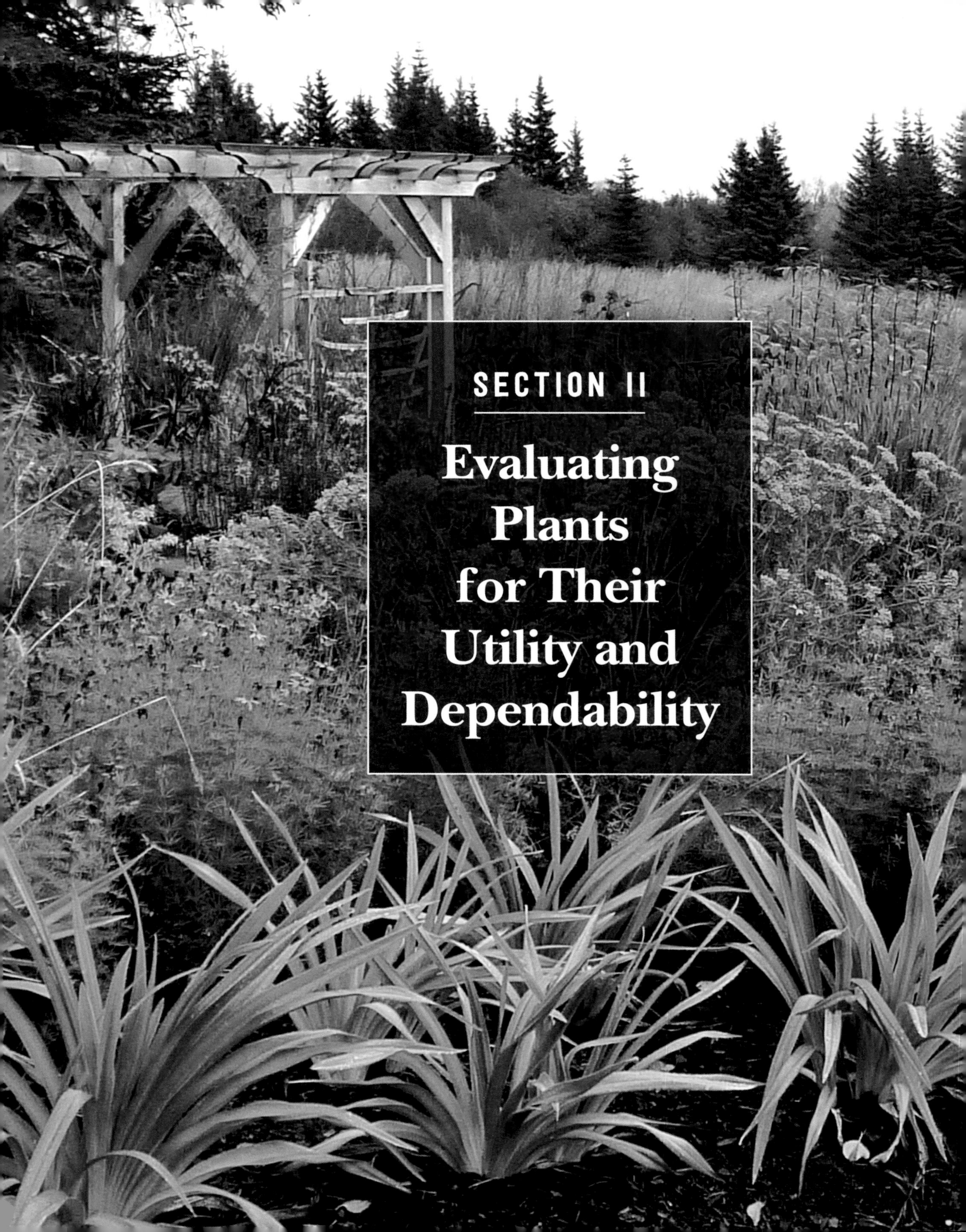

SECTION II

Evaluating Plants for Their Utility and Dependability

Besides artistic attributes and fragrance, there are additional qualities that you will find rewarding if you consider them when choosing new plants. Look for those plants that bloom for a long time, that change with the seasons—thus adding year-round beauty—or that are special in very early spring or late summer to fall. Give thought to a prospective plant's ultimate size and whether it will be in scale with the dimensions of your landscape. If one of your goals is having an easy-to-maintain garden, then well-behaved choices are a must. A new plant's horticultural needs should match the conditions of your environment so that it will remain healthy and vigorous without coddling. There may not be a single factor in this list that will prompt you to decide one way or the other, but a combination of characteristics might.

Let's consider each of these traits separately.

◂ The deep chocolate seed heads of threadleaf coreopsis (*Coreopsis verticillata* 'Zagreb') in the foreground, fading blossoms of yarrow (*Achillea millefolium* 'Paprika') behind them, and deep golden flowers of *Ligularia denticulata* 'Othello' near the pergola join tall grasses and other late bloomers to create a beautiful and colorful fall tableau at Stream Hill Park.

EARLY-SEASON VALUE AND FALL BEAUTY

◀ Ground covers, herbaceous perennials, and a dwarf shrub all contribute to this rich medley of fall color.

In much of Alaska and other northerly cold climates, the growing season is relatively short. As a result, plants that extend the season by revealing their beauty as soon as the snow melts and those that offer interesting displays beyond a killing frost are to be treasured. Spring bulbs and early natives can get the garden off to a colorful start well before the ground has warmed, whereas berries, fruit, and brilliantly colored fall foliage and bark can carry the show beyond the point when many herbaceous plants have crumpled to the ground. Plants that bloom during these shoulder seasons are particularly helpful to pollinators of all sorts, because they provide food sources when little else is available as well as when pollinators may need them the most—either as they recover from hibernation or a long migration or when preparing for winter.

It's not just color, however, that contributes beauty to early and late displays but also the subtle changes of the season. Spring's unwinding pinwheels of fiddlehead ferns, the swelling leaf buds on shrubs and trees, and the spears of new foliage pushing resolutely through the snow provide an anticipatory thrill as the long, dark winter melts away. As fall inevitably approaches, many plants don unusual and distinctive seed heads, often in the warm russet hues of autumn.

Removing the spent blossoms and their stems (deadheading) shortly after early-flowering plants finish blooming will often lead to a second flush of flowers in late summer or early fall. This is a satisfying way to increase the enjoyment you get from your garden.

Cherish late-blooming nectar-filled flowers, because they're capable of attracting bees and other pollinators to your garden when options for these helpful creatures are few and your work is finished. As the growing season wanes, take the time to relax in a comfortable garden chair and watch the bees swarm the remaining flowers; it is its own special experience.

▲ Lovely early-blooming native plants in the late Verna Pratt's garden include narcissus-flowered anemone (*Anemone narcissiflora*), fuchsia shooting stars (*Dodecatheon*), and pale blue forget-me-not (*Myosotis*). Verna's design.

▶ Long- and late-blooming sedum is dazzling in a light autumn snow.

▼ *Crocus vernus* 'Flower Record' blooms in very early spring.

◀ It's not just spring bulbs that provide nourishment to pollinators; those that bloom toward the end of the summer, such as drumstick allium (*Allium sphaerocephalon*), do too. Now that most of your chores are completed, sit back and enjoy the show!

▼ Early-blooming, bright yellow Leopard's bane (*Doronicum* 'Little Leo').

WINTER DISPLAY

Depending on what the average depth of snow is in your region, you may need to employ tall, woody plants, and especially trees, to provide visual gratification in the winter. Their strong silhouettes and powerful architecture rising above the snow provide dramatic testimony to the hidden presence of a garden beneath. If they have colorful or strongly textured bark or bright berries, all the better. Large shrubs with brilliantly colored stems are also prized for their addition to the winter landscape. Two hardy shrubs, red-twigged dogwood (*Cornus sericea* 'Cardinal')

◄ 'Miss Kim' lilac (*Syringa pubescens* subsp. *patula* 'Miss Kim') in the garden of Rita Jo and Leroy Shoultz. Photo courtesy of Rita Jo Shoultz; Rita Jo's design.

▲ LEFT TO RIGHT:

London pride saxifrage (*Saxifraga* x urbium 'Aurea punctata').

Red rose hips (*Rosa acicularis*) glisten in mid-winter.

Primula florindae seed heads hold "ice cream cones" of snow with a backdrop of feather reed grass (*Calamagrostis* x *acutiflora* 'Karl Foerster').

and yellow-twigged dogwood (*Cornus sericea* 'Flaviramea'), are excellent examples, with branches that glow against a blanket of snow.

Evergreen conifers, be they trees or shrubs, glistening in the low winter light can be simply stunning. The presence of their needles and cones throughout the winter adds true substance to the otherwise dormant landscape. I especially enjoy the appearance of a light layer of snow blanketing their branches.

Tall cool-season grasses, some of which can reach eight feet in height, will add a graceful elegance to your winter garden. Many of these will stand erect through all but the wettest snows. If your snow cover is fairly thin, you may be lucky enough to relish the foliage of evergreen perennials through much of winter as well.

BLOOM TIME AND LENGTH OF BLOOM

If you are trying to achieve certain color schemes, it's especially important to know when each plant will bloom. If you resort to a simple color scheme in which all hues work well together, the timing of flowering is much less important. On the other hand, the more fully you use all the attributes of great plants, the less important bloom time becomes, because the other strong characteristics of your plants will carry the day until the next flush of flowers appears.

Nonetheless, a long period of bloom is one of the very important qualities I've considered in identifying "cool" plants. The longer a plant blooms, the better! Most perennials offer their flowers for approximately two to three weeks during hot summers or three to four weeks during cooler ones, but a few special plants will put forth flowers for two months or longer. Among the longest-blooming plants we are able to grow in cold climates are two excellent cultivars of catmint, *Nepeta* x *faassenii* 'Walker's Low' and *Nepeta* x *faassenii* 'Six Hills Giant', as well as the many cultivars of masterwort (*Astrantia major*). Count these among your best supporting actors. They will aide in creating a colorful garden that will continue to perform during transitional periods and between bursts of color from your other plants.

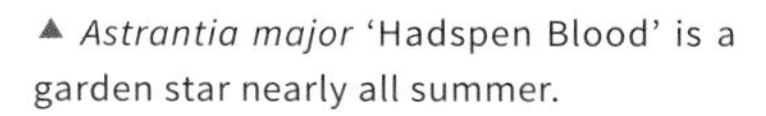

▲ *Astrantia major* 'Hadspen Blood' is a garden star nearly all summer.

▲▲ Columbines of all sorts bloom for an extended period (*Aquilegia* 'Colorado Violet and White').

► Fall-planted, summer-blooming drumstick allium (*Allium sphaerocephalon*) is radiant for nearly two months.

◄ Long-blooming drifts of yarrow (*Achillea millefolium* 'Terracotta'), orange avens (*Geum* 'Totally Tangerine'), golden grass (*Alopecurus pratensis* 'Aureovariegatus'), and lavender catmint (*Nepeta* x *faassenii* 'Walker's Low') make the Cathy and Scott Ulmer garden attractive and colorful for months on end.

ATTRACTIVENESS TO POLLINATORS

There are several components necessary for making a garden attractive to and supportive of pollinators. They include a food source, water, shelter, a place to raise their young, and the absence of pesticides. The most imperative of these is the last, the absence of pesticides. Because pesticides are typically indiscriminant, they can kill helpful insects as well as pests.

The flowers in our gardens provide nectar and pollen that sustain many pollinators, particularly bees, butterflies, and hummingbirds. The availability of food is especially vital during the earliest and latest periods of our growing season, as already discussed in "Early-Season Value and Fall Beauty." In addition, growing a robust variety of plants will attract a greater diversity of pollinators.

Aside from early and late bloomers, there are additional characteristics that make certain plants more attractive to pollinators than others. In general, butterflies are attracted to red, yellow, orange, pink, and purple blossoms but will visit flowers of other colors, too. Flat-topped flowers and those that are clustered offer a "landing zone" on which butterflies can perch as they feed, and you will often see them doing just that. However, because butterflies have a long proboscis, they can also reach nectar that lies deep within tubular-shaped flowers. In fact, they are uniquely suited to do this.

Though many may think only of the European honeybee when considering bees, in actuality about 4,000 species of bees are native to North America. They vary in size, color, nesting habits, and, to a lesser extent, pollinating techniques. With all this diversity, it's difficult to generalize, but everything I've read indicates that bees cannot see red. However, they do see in the ultraviolet range of the spectrum, which is invisible to humans. It might seem

◀ Giant yellow scabious (*Cephalaria tatarica*) is swarmed by bees when it blooms late in the season. Denice and Roger Clyne's garden.

For more detailed information about native bees and other pollinators, consult:

National Resources Conservation Service (NRCS)
www.nrcs.usda.gov/wps/portal/nrcs/main/national/plantsanimals/pollinate/

U.S. Fish and Wildlife Service
www.fws.gov/pollinators/Index.html

U.S. Department of Agriculture, Forest Service
www.fs.fed.us/wildflowers/pollinators/gardening.shtml

like magic to us, but to bees, many flowers have a bull's-eye, streaks, spots, or other patterns visible in the ultraviolet range that guide them to the pollen. Most bees are attracted to white, yellow, purple, or blue flowers, but it's not uncommon to see them on a wider range of hues. Single flowers have more pollen than doubles because the extra petals in doubles replace some of the pollen-laden anthers found in singles. This makes the pollen more easily accessed by bees, as does the configuration of flat or shallow blossoms such as daisies. Bees swarm plants in the mint family, like catmint, sage, and lavender, as well as those with hidden nectar spurs, including columbine, delphinium, monkshood, and beebalm. They also flock to plants rich in nectar, for example, the billowy peonies so treasured by brides.

Hummingbirds have no sense of smell, so they rely on sight to locate flowers that have the nectar they seek. While they feed on nectar, pollen can collect on their faces to be carried to the next flower. They are most strongly drawn to red and orange flowers. Vertical elements such as flowering trees and shrubs, vines, and hanging baskets will also attract hummingbirds, allowing them to seek shelter and offering a place for nests and high perches from which to survey "their" territory and to elude predators.

Tips for creating pollinator habitats:

Include plants for caterpillars, warming rocks or other flat basking places for butterflies to warm their wings, and a muddy location so that butterflies can "puddle."

Offer shallow water sources with an easy exit ramp for native bees so that they won't drown. Keep an area of open soil for those that build nests in the ground.

Leave those spider webs in place! Hummingbirds use a variety of materials to build their nests—lichen, twigs, pieces of leaves, grass plumes, and seed fluff—and hold it all together with spider silk. They prefer a trickle of water or mist rather than a birdbath.

▼ Bees visit peonies on a regular basis. Alaska Botanical Garden.

▲ Providing habitat for butterflies and moths is only half the equation; the other half is furnishing a food supply for their very hungry larvae. In this photo, taken by my daughter, Jocelyn Eisenlohr, in her garden, a Monarch caterpillar finishes off a milkweed (*Asclepias* spp.) that she planted just for him. In Alaska and the northern regions of Canada, native lupine is a wonderful food source for the butterflies that visit there.

▼ Stonecrop (*Sedum kamtschaticum*) with a color-coordinated visitor!

▲ Northern Jacob's ladder (*Polemonium boreale* 'San Juan Skies') hosts a busy bee.

SIZE AND SCALE

The size of a plant specified on a plant label is the *anticipated* height and width at maturity or, in the case of woody plants like trees and shrubs, the expected size at ten years of age. These stated dimensions assume an environment that meets the plant's horticultural needs. They also assume growing conditions similar to the location where the plant evolved, was developed, or was field trialed. In northern latitudes and cold climates, we often experience somewhat different results. Herbaceous perennials may exceed expectations because of the extremely long days, ample sunlight or, in some locales, frequent rainfall. Woody plants might lag behind expectations because of abbreviated northern growing seasons and their associated long, cold winters. However, the plant tag or other reference material can be used as an approximate guide until you learn more about how a particular plant performs in your unique environment.

▲ Allowing sufficient room for future plant growth minimizes future transplanting. Cathy and Scott Ulmer's garden.

▼ Large groupings of hardy geraniums (*Geranium* 'Sabani Blue'), painted daisies (*Tanacetum coccineum* 'Robinson's Red'), and purple dame's rocket (*Hesperis matronalis*) are in scale with the dimensions of a huge garden at Stream Hill Park.

Size is an important consideration, because you will want to place your plant in a location where it will have room to develop without overcrowding its neighbors, structures, or pathways. Proper placement will help you avoid the need to transplant in the future. Expected size will also help to determine where in your garden to place each plant relative to others so that it neither obscures the view of a plant behind it nor seems out of balance with those nearby. If you want to create a layered, lush-looking garden, you can capitalize on the range of plant sizes to achieve your goals. In addition, you will bring about a more attractive result if your plants are in scale with the size of your garden and your home.

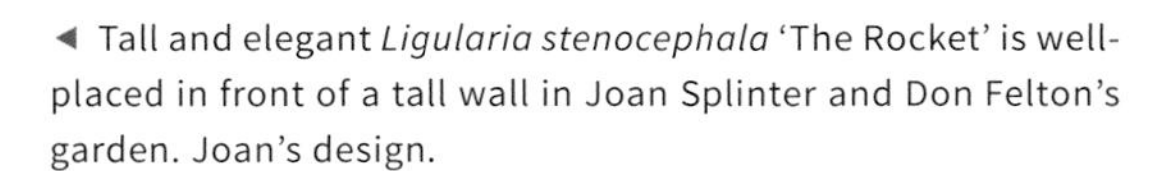

◀ Tall and elegant *Ligularia stenocephala* 'The Rocket' is well-placed in front of a tall wall in Joan Splinter and Don Felton's garden. Joan's design.

BEHAVIOR IN THE GARDEN

If you wish to have an easy-care, low-maintenance garden, it's paramount that your plant choices reflect that desire. Select plants that will be well behaved, ones that grow larger over time *in the same place where they are planted*, and not those that will wantonly toss viable seeds all over or that spread rapidly via underground rhizomes or root suckers; these characteristics are the antithesis of good behavior. Rapidly spreading plants require constant management and effort to keep them in bounds. That's not low-maintenance!

There's a large group of plants commonly referred to as "pass-around plants" because they are freely passed from one gardener to another—often because the generous gardener has "plenty to share." Beware of these well-intended gifts; there's a reason why the person handing you the plant has plenty to share!

When I first moved to Alaska and embarked on my Alaska gardening adventure, our dear homesteader neighbors gave me some of their plants. Among them were pale yellow butter and eggs (*Linaria vulgaris*), giant blue bachelor's buttons (*Centaurea montana*), and white oxeye daisies (*Leucanthemum vulgare*). At the time, I knew very little about any of these plants, as none of them were grown where I'd gardened before. That sweet little *Linaria* became a nightmare, marauding throughout my first garden within two seasons. It looked like a pretty and innocent miniature snapdragon. Instead, it was an absolute scamp! Shame on me for not learning more before planting it! I still keep one of the *Centaurea montana* plants because I love it as a cut flower, but I must do battle with it every spring, digging up all the seedlings it has propagated. Although I kept that one, I quickly realized that the other two had to go completely. Unfortunately, that was easier said than done. Twenty years after digging them up—and, I thought, totally eradicating them—I still find these aggressive interlopers popping up in random places throughout the garden.

Nearly all the many great plants suggested here have been very well behaved in my experiences with them; the few exceptions are clearly noted

◀ Beautiful ribbon grass (*Phalaris arundinacea* 'Picta') is a cultivar of a very aggressive grass known as reed canarygrass that is categorized as invasive in many locales. Even this cultivar is very tough to keep in bounds—it is *not* consistent with good garden behavior or low maintenance.

within their descriptions. To learn more about how each performs in other locations, I've consulted with respected, well-known gardening professionals from the far corners of Alaska and other cold climates and have used their feedback in making these recommendations. It is possible, however, that significant differences in climatic or growing environments may produce different results for you. For example, those of you who garden in generally cold climates but whose summers are very hot may find some of these selections to be too vigorous. On the other hand, a few of them, notably the Himalayan blue poppy (*Meconopsis betonicifolia*), will struggle in heat or humidity.

◀ Some of the most beloved plants can be real management headaches. Delightful daisies, so pretty in a vase or a wild meadow, create an enormous number of seedlings that may be unwanted in your garden. Jody Murdock's former Aurora Gems garden.

▲ Prickly rose *(Rosa acicularis)*, an Alaska native, has nice green foliage, lovely pink flowers that are followed in fall by brilliant red hips, but it suckers unmercifully making it difficult to keep in bounds.

VIGOR AND SUITABILITY

If a plant is beautiful and well behaved and has many other wonderful characteristics but struggles through a long, cold winter, it won't be a good choice. Nor will it be low-maintenance, because you will spend an inordinate amount of time coddling it to try to keep it alive. Selections that grow vigorously in your climate and conditions without being overly aggressive are ideal. They'll look and be healthy. Moreover, their natural defenses will better protect them from disease and pests. In short, they'll make your gardening life much easier and infinitely more rewarding.

◄ Hardy, sun-loving, and drought-tolerant plants look vigorous and healthy in an environment that perfectly suits their needs. Elise and Jay Boyer's sunny garden.

To be well-suited to its location in your garden, a plant must have its horticultural needs met. In the next section, I'll talk about horticultural requirements in more depth, but the basics include sufficient hardiness, the proper amount of sunlight and moisture, and a good match with your soil.

◄ These plants enjoy more shade, a cool summer environment, and a reasonable amount of moisture. Though perfectly suited to this shady coastal Alaska garden, Japanese coltsfoot (*Petasites japonicus* var. giganteus 'Variegata'), the large round-leafed plant in the middle ground, is likely to be too vigorous in a hotter summer climate or even a sunnier one in coastal Alaska. Elise and Jay Boyer's part shade garden.

EASE OF MAINTENANCE

I've already touched on two issues that contribute to ease of maintenance—plant behavior and suitability to your environment—but there are several other factors that you should also consider. How frequently must a plant be divided to maintain its vigor? Does it need to be deadheaded to look good or to keep it from indiscriminately reseeding? Does it require staking or other support? Is pruning required to keep it in bounds or needed to improve vitality? If so, how often? Is it susceptible to diseases common in your area? Does it attract destructive insects? Do the local mammals find it delectable? All these questions correlate to tasks that a gardener must perform to keep certain plants looking their best and thriving. Although you might so cherish a certain plant that you gladly carry out the tasks necessary to provide the care it needs in order to do well, an entire garden filled with prima donnas will consume a great deal of your time.

Creating a garden that is easy to maintain requires a focus on plants that seldom need dividing and that don't need deadheading, staking, or pruning. Varieties bred to be disease-resistant are the most desirable, as are those that are unpalatable to your local mammal population.

◄ Many gardeners decide to include tall and stately delphiniums in their gardens because they "love them" and thus are willing to tolerate their somewhat short-life and need for staking.

▲ Although the foliage of lady's mantle (*Alchemilla mollis*) is incredibly beautiful, especially after a morning dew or rain as seen here, its propensity to create many seedlings unless deadheaded regularly may be too much of a maintenance burden for some gardeners.

◄ A broad range of succulents and other drought-tolerant plants, if situated in a sunny, well-drained location, require virtually no care to stay healthy and beautiful. Denice and Roger Clyne's garden.

DO YOU LOVE IT?

ne of the least discussed aspects of a plant is how *you* respond to it. Do you love it or at least *really* like it? This is important! Do you adore the color of the flowers or foliage? Does the fragrance bring back fond memories? Do you find the texture, architectural form, or seed heads intriguing? If it's already in your garden, are you excited to see it return each year? To me, this is a crucial test. Since you will live with your garden every day, it should be filled with plants that you fancy unconditionally and utterly devoid of those that annoy you. After all, the purpose of your garden is to bring you enjoyment.

I seek plants that will likely become wonderful old friends: the ones that bring a smile to my face when they emerge to reveal their beauty, that look fabulous with their neighbors, or that extend a theme in my design. If they do not do these things, they might find themselves being taken for a "shovel ride", but not necessarily to the compost pile. Often a plant can be irritating simply because it is in the wrong place. For example, its color might be jarring with a neighboring plant. If that's the case but you like it otherwise, move it. But if you simply do not like the plant, then get rid of it!

◀ Your garden should be filled with plants that bring a smile to your face when they return each spring in the same way that our lab, Sunny, brings a smile to mine every morning.

SECTION III

Determining Whether a Plant Will Thrive in Your Garden

For many of us, our earliest attempts at gardening are frustrating. We decide to create a garden, so we buy a few pretty plants, dig some holes, pop the plants out of their containers, put them into the holes, and call it good! Although we water these new arrivals carefully and religiously, too often they struggle and ultimately fade away and die. Sadly, that's the end of the gardening story for some, but the more determined and tenacious among us search out the reasons why things went awry, and, more importantly, how to be successful in the future. Critical to this quest is learning to evaluate our growing environment and to prepare it optimally for plants. We also must recognize that different plants require different conditions to thrive. Finally, we must be able to align the two.

Let's look at these essential elements-for-success in more detail.

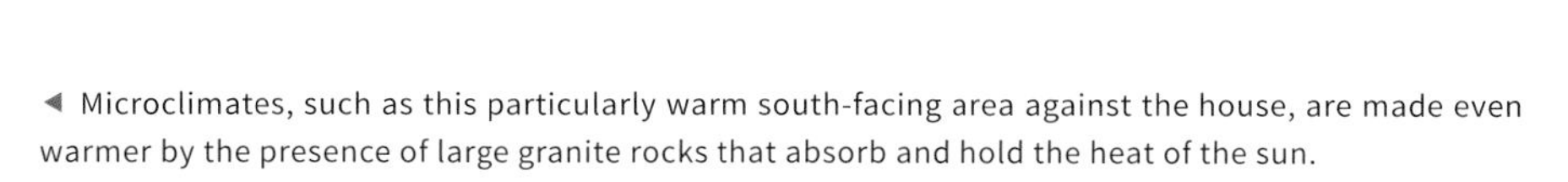

◄ Microclimates, such as this particularly warm south-facing area against the house, are made even warmer by the presence of large granite rocks that absorb and hold the heat of the sun.

► Notice the robust clump of lavender, a plant that normally struggles to survive in Zone 3 Alaska, looks hail and healthy in this "toasty" niche. Joan Splinter and Don Felton's garden; Joan's design.

UNDERSTANDING YOUR GROWING ENVIRONMENT

In order to determine whether a specific plant will do well in your garden, you must first observe, test, and understand the conditions of your growing environment. For example, are the areas where you have or will have gardens shady, or are they sunny or partly so? Does the amount of light vary throughout the day? How about throughout the season? This is easy to observe on any given day, but it takes a little discipline to pay attention to those changes and record them over several months. Doing so is, however, worth your time since you will need to accomplish this task during only one season, while the benefits of site-appropriate plant selection and placement will last for the life of your garden.

One of the most important tenets of successful gardening says that you should use plants that are well matched to the soil. To achieve this in your garden, you must learn about your soil. What is its physical makeup? Does it drain well, or does it hold moisture? Is it primarily sand, heavy clay, or a desirable loam? Is it acidic or alkaline? Understanding the answers to these questions will give you the insight you need to select plants that will do well in your garden's soil.

The average extreme low temperatures during the dormant part of the year and extreme high temperatures throughout the growing season determine your plant hardiness zone and heat zone, respectively. Other factors that help decide a plant's suitability for your garden include frequency and severity of windy conditions, exposure to salt through ocean spray or road maintenance, and amount of foot traffic in the area.

These conditions can vary considerably even within a small garden. Additionally, structures made of masonry, like retaining walls or your home's foundation and wind breaks such as those created by a stand of trees, large shrubs, or a nearby building can create a microclimate that is warmer or more sheltered than other nearby areas. Conversely, fences and buildings sometimes create cooler, shaded areas on their north side. As you grow to understand the details of your property, you can take advantage of these nuances and expand the range of plants that will thrive for you.

Testing your soil can bring huge advantages

Soil testing is a simple and fairly inexpensive process, but one that results in a lot of useful, even critical, information.

Look for Cooperative Extension Publications with Guidance on

- **Soil sampling**
- **Selecting a soil test laboratory**
- **Soil and fertilizer management**

CULTURAL REQUIREMENTS

As plants have evolved in their natural habitats, they have adapted, or developed strategies for succeeding in the environmental circumstances prevalent in their indigenous region. A plant's success in your garden will depend on how well matched your garden is to the habitat where the plant evolved, especially in the amount of sunlight and moisture, soil type, soil pH, soil fertility, and temperature range. Other factors, such as wind exposure, humidity, and day and season lengths, are also important. Taken together, these are known as a plant's cultural needs or requirements. If we can understand the environment in which a plant evolved and flourished and then place that plant in similar circumstances, we increase its potential for success. One of the reasons why plants that are native to a specific area are so successful when invited into a local garden is that they have evolved in and adapted to that area.

Fortunately, many plants will tolerate a range of conditions that vary somewhat from their native habitat, allowing us to enjoy a fairly wide range of selections beyond local natives. There are regions around the globe, especially in the northern latitudes, that may have habitats similar to your own and from which some wonderful plants hail. But to make good plant choices for your landscape, you'll need to understand more about cultural specifics.

Most cultural needs are expressed as a range or a guideline on a scale. Sunlight is a measure of how many hours of sun a plant needs per day. You'll see it described as *full sun* (six to eight or more hours of direct sun), *part sun* or *part shade* (up to six hours of direct sun, four of them in the morning), *dappled shade* (sunlight filtered through the foliage of taller plants and trees), and *full shade* (fewer than three hours of sun, early in the day). Full shade is generally found on the north side of a building or fence or under a canopy of dense trees such as spruce. These areas can be challenging places for plants in cold climates. Without the benefit of warming sunshine, the soil will stay cold remarkably far into the season, dramatically shortening the growing period in that area.

In truly cold climates, for a perennial plant, tree, or shrub to survive and do well through the winter, it must be "hardy" enough for its location.

◄ Not only is the north side of a fence or building largely shaded, but in cold northern climates the soil will stay cold there well into the season, making careful plant selection even more essential for success. Kathy and Mike Pate's garden.

Experiments in Pushing your Zone

Invariably we find a "must-have" plant that lies *just* outside our hardiness zone. To increase your odds of success when experimenting with a plant in this category, you can do the following:

- Cover it with winter mulch *after* the ground has frozen to protect the crown and roots.
- Site it in the most advantageous location for all its other horticultural needs.
- Plant it near the foundation of your home so that it benefits from the heat that escapes from your home during winter. (Be careful, though, because sometimes this backfires when the plant breaks dormancy during a warm spell in winter.)
- Give it the best drainage possible to minimize the stresses of freeze–thaw cycles.
- Cross your fingers!

Hardiness is a *guideline* for how much cold a plant can tolerate. To express this easily, the U.S. Department of Agriculture and the Canadian Department of Natural Resources have established and mapped a set of standard "hardiness zones." Although the USDA zone map takes into account only temperature, its Canadian counterpart also considers snow cover, wind, and rain, as well as the average length of the frost-free period. As an Alaska gardener, I know the importance of these additional factors in a harsh climate and would love to see the USDA take them into consideration as well.

The American Horticultural Society has developed a heat zone map, the counterpart to the hardiness maps. Although not as broadly referenced as the hardiness guidelines, it does reflect the importance of summer heat. Some plants that are hardy in extreme cold are not well suited to extreme heat in summer, while other hardy species might languish in a cool summer. For example, the beautiful and sought-after Himalayan blue poppy, *Meconopsis betonicifolia*, is wonderfully hardy but becomes bedraggled in climates that have hot summers. Conversely, two of my favorites from my Pennsylvania childhood, purple coneflowers (*Echinacea purpurea*) and black-eyed Susans (*Rudbeckia fulgida*), fall into the latter category. Try as I might to ignore reality, neither of these heat lovers ever gets around to blooming in coastal Alaska; both eventually succumb to the unsuitable environment and disappear from my garden.

Hardiness zone is a measure of average annual minimum temperatures in a given area. The range of zones in which a plant is expected to succeed is the key to how hardy a plant is; the lower the number, the more cold-tolerant it will be. Simply speaking, in a northern climate hardiness measures how much cold a plant can tolerate without protection *provided that its other cultural needs are met.* For example, cold drying winds, insufficient sunlight, or a poor soil match can reduce a plant's ability to thrive within its identified zonal range. On the other hand, winter mulch or dependable snow cover can protect a plant so much that it will withstand colder temperatures than it could tolerate otherwise.

Much of interior Alaska and northern Canada fall within Zone 1, progressing to Zone 2 as you move closer to the ocean or farther south. Along the coasts, zones vary from 3 to 5, with some areas in Southeast Alaska designated Zone 6 or 7 on the USDA maps. The two southern coasts of Canada are also rated Zone 6 or 7. Toasty! The upper Midwest of the United States and much of New England fall within Zones 2–5, depending on location. Zones 1–5 are all considered cold climates. These maps don't take microclimates or detailed local variances into account, but they do provide a useful starting point.

If you are just beginning your gardening experience, I suggest taking a conservative approach to zones. In other words, if the map says your area is a Zone 3, select plants that are rated for Zone 3 or lower. Once you become more experienced you may want to experiment, a few plants at a time, to see how far you can push the zone limit by recognizing and making use of microclimates or by taking advantage of the benefits of winter mulch or consistent snow cover.

One of the most important cultural requirements for plant health is a good plant–soil match, yet this issue is frequently overlooked perhaps out of a belief that the concept is too complex to grasp. Not true! We're about to cover it in just one page.

Soil is the medium through which plants receive their sustenance, so its suitability for a given plant is vital for gardening success. Because dramatically or quickly changing the nature of soil is difficult, you'll achieve success much more easily by learning about the soil you already have and selecting plants that will thrive in it. Experience has taught me that selecting plants first, then trying to change the soil to meet their needs is an approach fraught with disappointment.

The metrics associated with soil suitability include fertility, pH, the physical make-up of the soil, and its ability to retain water. Fertility measures the amount and kind of nutrients present in the soil. Soil pH, a measure of the soil's relative acidity or alkalinity, is expressed as a number ranging from 0 (extremely acidic) to 14 (extremely alkaline), with 7 representing neutral. Physically, soil can be made up of small particles (clay) as well as large ones (sand) and those of intermediate size (silt). Loam, an approximately equal mix of clay, sand, and silt, is a gardener's ideal, because it is a good balance between fast- and slow-draining soils. The amount of decayed vegetative matter (compost) as well as the particle size will also affect the water retention capacity of soil. Soils with a high concentration of clay are extremely water-retentive, whereas sandy soils drain quickly. Ironically, the addition of compost to clay soil results in better drainage, while adding it to sandy soil produces better moisture retention.

The most dependable way to get an accurate understanding of the fertility, pH, and physical makeup of your soil, as well as how to improve it, is to have it tested. I strongly recommend that you do so. Believe me, beginning with this step will ultimately put you dollars ahead and help avoid disappointment later. To learn more about where and how to have your soil tested a good place to start in the United States is your local Cooperative Extension Service office. In Canada, contact your provincial Agricultural Extension. Many private labs also do soil testing; an Internet search will provide a long list of options. Be sure to ask whether the lab you are considering conducts tests appropriate for your location.

The moisture needs of plants are often described with phrases such as *drought-tolerant*, *requires moist, well-drained soil*, or *requires moist soil.*

A drought-tolerant plant, once established, can survive extended dry periods and, in some cases, even desiccation. Many plants that appear to be gray because of hairs on their leaves, most succulents, and plants with fleshy roots, including some bulbs, will tolerate drought once established. It is worth noting, however, that even plants of this type require moisture until they have established an ample root system. This will normally take from one to two seasons, depending on your particular environment and the type of plant.

A large majority of garden plants are described as requiring moist, well-drained soil. Although this combination might sound paradoxical, what it describes is a soil that drains *excess* water away fairly quickly, staying damp without being soggy, much like a well-squeezed sponge.

Moist soils are those that retain moisture for an extensive period of time. Certainly there are different degrees of moist soil, ranging from damp to downright boggy. Fortunately, for folks who have extremely water-retentive soil, there are plants that thrive in that environment, notably a group referred to as "marginals", because they grow naturally at the margins of water, along streams and at the edges of ponds.

After you understand the parameters of your environment, focus on plants that will thrive in what you have. To assist you, the plant descriptions in this book include a quick reference that describes the horticultural needs of each plant.

◀ Even drought-tolerant plants like *Lewisia tweedyi* with its long taproot, and succulents, such as this intriguing-looking *Sedum*, need moisture until they are well established.

SECTION IV

At the Nursery

Nurseries are intoxicating places, especially in early spring, when they are packed with new offerings for the season. This is especially true for cold-climate gardeners, who wait stoically through interminable winters for the arrival of long days and increasing temperatures. The smell of warm soil, the humid, fragrance-laden air, and even the droplets that splash on your head as you walk under freshly-watered hanging baskets are all part of the thrill and anticipation of what's soon to come.

Each spring, I make a point of visiting my favorite local nursery on the day it opens. I'm eager to see all the new plants, but I also know that the day will be a social one, since so many of my gardening friends will be there too. We must be an amusing sight as we dash about, calling each other to come see this or that "really cool" plant. Then we stand around it and fantasize about where we might find room for it in our already packed gardens. No matter, we'll find a spot! There's always room for one more plant—or so we convince ourselves as we take a flat full of "one mores" to the checkout table. Oh, what a joyous day!

One of the challenges for cold-climate gardeners is dealing with the vagaries of early-season temperature swings. If we purchase our new treasures before outside temperatures are reliably above freezing, then we need a warm place to keep them while we wait. A heated

◄ There is nothing like the first day at the nursery after the long winter wait! Herbaceous perennials line the tables at Alaska Hardy Gardens in Homer, Alaska, during the second week of April.

▶ Colorful mixes of perennials, exotics, and annuals abound at The Plant Kingdom Greenhouse and Nursery, Inc. in Fairbanks, Alaska, during the first week of May.

greenhouse, perhaps a hoop house (also called a high tunnel), or even a warm room in the house that lets in ample light may suffice. Some nurseries will keep plants for us briefly for a nominal fee. This is a real plus for gardeners who don't have protective structures of their own.

Ultimately, though, plants will need to transition from one of these protected environments to the great big outdoors, with its bright sunshine, erratic temperature swings, and sometimes windy conditions. The process used to successfully guide plants through this is called "hardening off."

One way to do this is to move your plants outside for a brief period of time each day, increasing the exposure duration steadily, and then return them to their protected location until the next day. After a few years of using this technique, I decided it was too much work. Worse, I sometimes got involved in other activities and forgot my tender little charges. Not a happy result!

Here's an easier technique I learned from a friend. Move plants outside onto a table or platform to keep them off the bitterly cold ground, hydrate them well taking care not to wet the foliage, and then cover them with three layers of floating row cover (Reemay is one brand.) After four or five days, check the moisture level, and then remove a layer of fabric. Repeat this process four or five days later. Leave the last layer on the plants until you are sure that nighttime temperatures will not fall below freezing. When you remove it, the plants will be hardened off *and* conditioned to the sun.

Geranium,
Zonal
Maestro
Light Pink Parfait
$5.95
12-16"
Sun

$ 89 95

CHOOSING A NURSERY
IF YOU HAVE A CHOICE

Like anything else in life, some nurseries are better than others. There are a number of factors that make the best of them stand out. To me, the first criteria must be the health and vigor of a nursery's plants. If the plants aren't well tended, especially as the season evolves and pest insects become an issue, things you buy there might introduce problems into your garden, certainly not an experience you need. A clean, orderly, well-ventilated environment is a must. I expect plants to be correctly labeled, well presented, and properly hydrated—neither too dry nor too soggy—and I strongly prefer nurseries that employ integrated pest management practices, using beneficial insects rather than chemicals to control troublesome insects. It might be obvious, but to be thorough, let me add that pots should be weed-free.

Beyond these basic health issues, I look for nurseries that have a large variety of plants that are zonally appropriate for the community they serve. Employees should know what's in inventory and where to find it. The reality is that few employees will have in-depth knowledge of each variety of plant being offered, so good signage, a stack of reference books for customers to use, and at least one knowledgeable senior staff member or owner are extremely desirable, if not necessary.

Though it's the least decisive element in the equation in the long run, competitive pricing—not necessarily the lowest price in town, however—is always a plus.

◀ Healthy, well-hydrated, pest-free plants are a must when choosing a nursery. Container designs by Cyndie Warbelow, proprietor of The Plant Kingdom Greenhouse and Nursery, Inc. in Fairbanks, Alaska.

▶ The more varieties, the better! Alaska Hardy Gardens, Homer, Alaska.

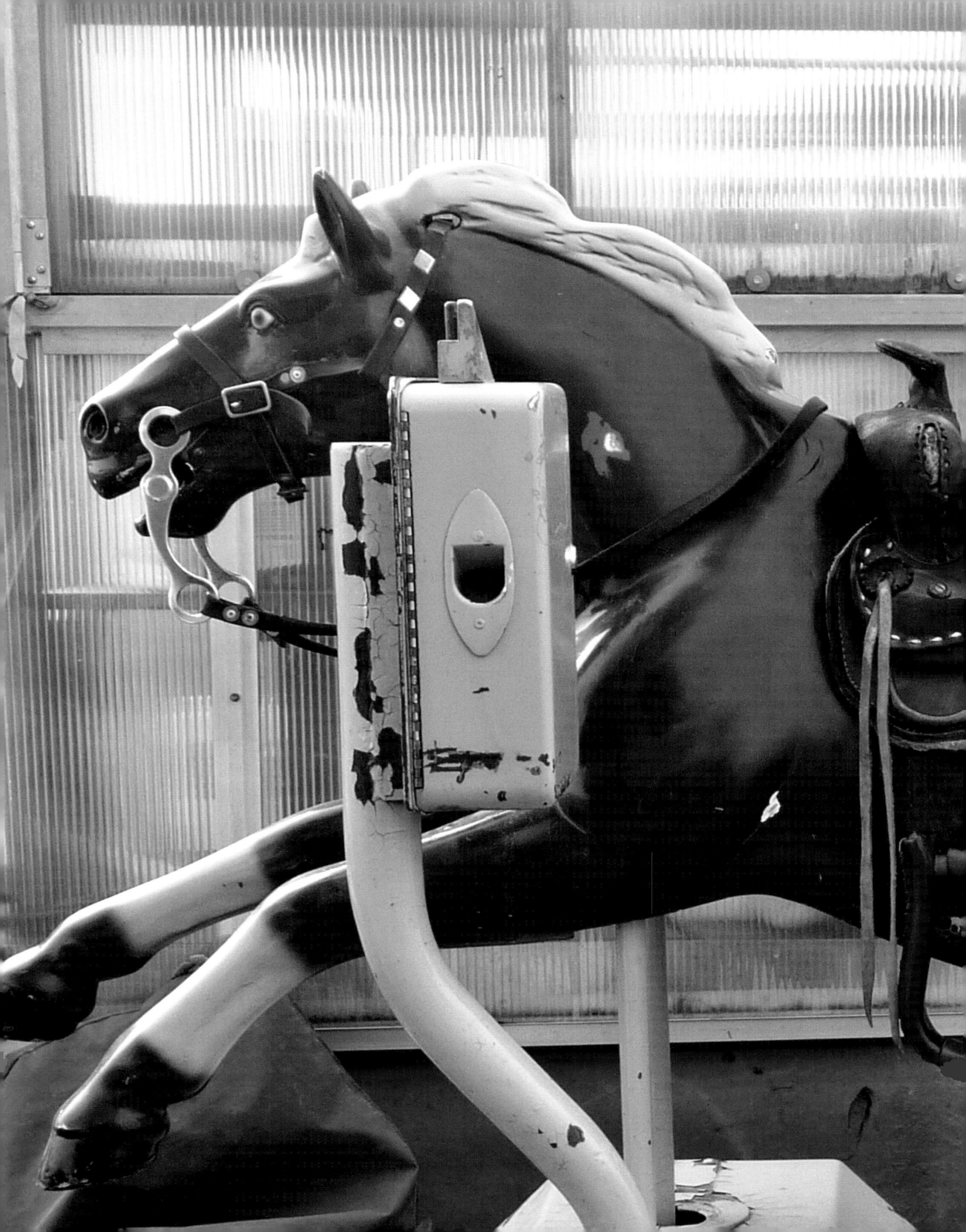

LOCAL VERSUS ONLINE

n many areas of our lives, online shopping has become the norm. For those of us in remote areas of Alaska, it is often the only avenue we have to readily acquire certain items. Fortunately, as Alaskan gardeners, we are blessed with a wide variety of nurseries in many different locations; I hope you are, too. Although I often buy seeds and bulbs online, only an extremely rare situation would prompt me to order live plants in this manner.

First, as I've mentioned, I enjoy visiting my local nurseries. More importantly, though, I want to see and inspect a plant before I purchase it. It's important to me to select the "best of the best" of the options, and for those plants to be as fresh as possible. If you have had good luck with specific online nurseries and are happy with their products, then by all means continue on that course. On the other hand, if your local nursery is offering high-quality plants, I encourage you to support it with your business.

◀ In addition to allowing you to carefully inspect potential purchases, nurseries can provide special and often surprising experiences that you just can't get online. Ann's Greenhouse, Fairbanks, Alaska.

LILY LOOKS

SELECTING THE "BEST OF THE BEST" PLANTS FOR YOUR GARDEN

Not only are different kinds of plants not all created equal, but neither are all specimens of the same type equal. Sometimes, while looking over several flats of a given plant in a nursery, it's hard to see a difference among the plants. That's ideal, and it should be the nursery's goal; however, it's often an elusive one. Usually you will find that some plants will be a little bigger or smaller than the average in the flat; similarly, a few might be in bud, while others are in full, gorgeous bloom. The question is—which are the best to buy?

Because cold climates can be somewhat challenging to new plants, it's best to select the largest plants you can afford. If the nursery offers the same cultivar of perennial in four-inch and six-inch pots, purchase the six-inch size. If a given shrub is available in one-gallon and five-gallon pots, take the five-gallon size. The larger the pot and plant, the more mature the plant should be. Purchasing mature, healthy plants will increase your success rate considerably in our challenging environments and is worth the increase in price.

Choosing the plants that are larger than average might seem like an obvious strategy, and in most cases, it is a good one. The exception, however, is when a plant is so big that it is straining at the confines of its pot. This often produces a root-bound condition in which roots circle the pot repeatedly, making a dense, tangled mass. If not corrected, this condition eventually leads to the roots "strangling" themselves, ultimately causing the demise of the plant. If a plant looks great otherwise but you suspect that it has been in its pot too long, carefully slip it out and look at the roots. If the plant is seriously root-bound, keep looking for a better option.

When it comes to making your purchase, plants that are in bud are a better choice than those that are in bloom. Although it's inviting to see one blooming at the nursery so that you can confirm that the flower color is what

▲ All else being equal, select the largest plants you can afford. This is especially important when selecting slow-growing woody plants such as this 'Fat Albert' Colorado spruce (*Picea pungens* 'Fat Albert'). Alaska Hardy Gardens.

◄ Plants in bud are a better choice than those in bloom. Alaska Hardy Gardens.

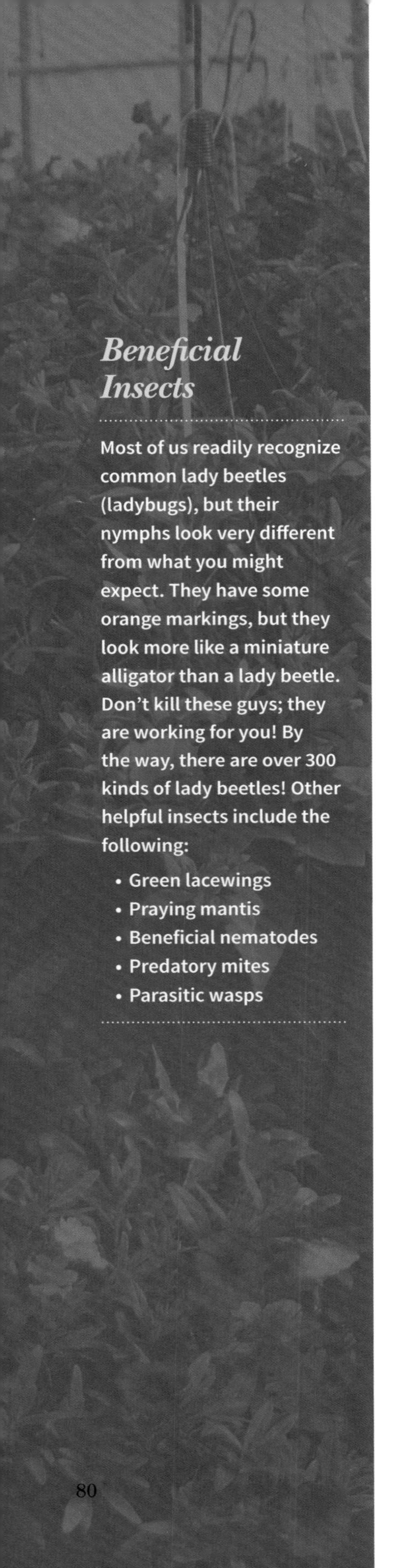

Beneficial Insects

Most of us readily recognize common lady beetles (ladybugs), but their nymphs look very different from what you might expect. They have some orange markings, but they look more like a miniature alligator than a lady beetle. Don't kill these guys; they are working for you! By the way, there are over 300 kinds of lady beetles! Other helpful insects include the following:

- Green lacewings
- Praying mantis
- Beneficial nematodes
- Predatory mites
- Parasitic wasps

you desire, buy the ones that are not yet blooming. It's important for a new plant to establish a healthy root system in your garden during its first growing season, because this will improve the odds that it will survive the first harsh winter in a cold climate. If the new plant is expending too much energy on supporting a full cover of blossoms, little will be left for critical early root development. Some plants benefit immensely from actually cutting away the bloom stalks before the flowers open the first year. That's a hard thing to bring yourself to do, I know, so at the very least, select plants that will have a longer window of time during which to put out new roots before their bloom cycle begins. Similarly, I don't let new perennial plants go to seed in their first season, because this also uses a lot of the plant's resources that are better directed to the roots. Finally, from a purely selfish point of view, it's better to select plants that are not in bloom so that *you* can enjoy the flowers in *your* garden rather than have everyone else delight in them at the nursery.

When selecting the best of the best, plant health is critical. Careful inspection of the foliage can provide many clues to the well-being of a given specimen. Look for foliage that appears fresh and turgid, neither droopy from too little water nor mildewed by too much moisture. Leaves should be blemish-free and richly colored. They should not be yellowed if they are supposed to be green, nor pale yellow when they should be gold.

Fresh new growth is always a positive indicator, because it is a clear sign that the plant is actively growing and is not in shock from being transplanted or shipped. If a plant has multiple stems emanating from the center of the crown, a somewhat mounded area just above soil level, then look for new growth in that area. If the plant has a branched structure, then look for new shoots along or at the ends of its branches.

Inspect plants for damaging insects. Avoid plants infested with aphids, spider mites, white flies, or other parasites. Not only do these insects weaken the plant, but they can also spread to other plants in your garden or greenhouse when you take them home. Many nurseries practice integrated pest management (IPM) using beneficial insects rather than sprays to control destructive insects, especially inside their greenhouses. Therefore, you might see lady beetles, lacewings, praying mantises, and parasitic wasps or their larvae on the nursery's offerings. If you see bugs and are not sure whether they are pests or beneficial insects, ask someone who works for the nursery to help you identify them. If they are a beneficial insect, you'll learn something; if they are a pest, you will have helped your nursery head off a more serious infestation. Don't be shy about this issue. The nursery folks will appreciate your keen eye if they have a developing problem.

Some nurseries sell beneficial insects for you to release into your own greenhouse. If you have this option, then I recommend that you do so as a preventative step. In other words, release beneficial insects *before* you see pests. Certainly if you do discover pests among your plants, you should act quickly and inundate them with beneficial insects.

Another pest you should look for is the ubiquitous slug. On occasion you'll spot one of these on the bottom of a pot that has been sitting on the ground. More often you'll see only the damage on the leaves of the plant; slugs make round holes in foliage. Bypass those specimens that have obvious slug damage; the soil in the pot might contain slugs, which are easy enough to spot, but also slug eggs, which are just as easy to miss. This is a pest that you absolutely do not want introduced into your garden.

Weeds are another problem that you don't want to bring home from the nursery, be they indigenous or imported from afar. Examine the surface of the soil in each pot you are considering. If you see weeds growing in it, you might want to reconsider and select a different plant. Unfortunately, some potting mixes, as well as the soil that comes with field-grown plants, can be a source of new weeds—for your garden but also for your state or territory. Use caution to avoid introducing new problems into your life.

◄ Fresh-looking blemish- and pest-free foliage is an excellent indicator of plant health. Alaska Hardy Gardens.

◄ Whereas one of the goals of a nursery should be to present uniformly developed plants, your goal is to find the one that's just a little bit better than the rest. Alaska Hardy Gardens.

Keeping Out New Weeds

As discussed at some length in *There's a Moose in My Garden,* you can minimize the introduction of new weeds into your garden by carefully and consistently removing soil from both the top and the bottom of pots in which your new plants arrive. The top three-eighths of an inch is the most likely place for weed seeds to lurk, while the bottom half-inch may harbor slugs or their eggs. Remove this soil and discard it in a landfill or other trash receptacle. Do not add it to your compost pile.

OTHER SHOPPING TIPS

Although all plants you buy should be attractive, well-formed, healthy, and pest-free, each major category of plants has some additional or unique characteristics to consider.

◀ Mixing purple, pink, and lime green together is not something I might have tried in my perennial gardens until I learned how much I liked the combination in this container.

Annuals

Because many annuals will bloom nearly all summer if you deadhead them regularly, it is not as critical—though it is still recommended for maximizing your enjoyment—that you choose plants in bud versus those in bloom. Some nurseries do an excellent job of pinching back annuals as they begin to grow in their pots. This results in more stems and a fuller plant, and thus more blooms. So look for plants that have been pinched and are well branched and full.

You will find that some annuals, for example pansies, do better during cool weather, whereas others, like verbena, are at their best in baking heat. In either event, I find annuals a perfect vehicle to try different color combinations and to indulge my passion for something wild and different.

▲ Well-pinched annuals await shoppers at Forget-Me-Not Nursery in Indian, Alaska. It's nice to see one in bloom to confirm the color.

Bulbs

Bulbs should be firm and free of bruises, cuts, and signs of disease. The papery covering or "tunic" found on many bulbs might be loose or even partially or totally missing. Select those that have the most intact tunic, because it provides a layer of protection for the bulb. The larger the bulb within its peer group, the better. When selecting bulbs that naturalize, returning year after year in larger numbers, the size is even more important, because the larger the bulb, the sooner it will divide and increase the number of blossoms that it can produce. When deciding where to plant bulbs, always select a well-drained location. This is particularly important if your area is subjected to freeze–thaw cycles.

Some bulbs, though they produce beautiful blooms, are susceptible to foraging by burrowing animals such as shrews and voles or to being decapitated by browsing moose or deer. Sweet-tasting tulips are the classic example.

Thoughts on Deadheading

First, for best results, remove the spent flower *and* its stem. Cutting the flower stalk to its base or at least down into the surrounding foliage will give you a much tidier look than removing the flower only. The latter can leave an unattractive empty stem sticking out of the foliage.

Deadheading will prolong and enhance the bloom display of flowering plants by reducing the competition for plant resources that seed development causes. It also prevents unwanted seedlings.

But what if you want both prolonged bloom and seedlings? Then deadhead throughout the early to middle part of your season. As the season winds down, let some of your desirable self-sowers go to seed to provide a crop of new plants next year.

▲ As is the case with a number of herbaceous perennials, some annuals do not need deadheading to continue their bloom. This is heaven for a busy gardener.

Burrowers love the bulbs; browsers can't refrain from munching on the flowers. This is unfortunate, because tulips are incredibly beautiful and are available in such a glorious array of colors. If you know you have garden residents that will feast on edible bulbs, you might want to select bulbs like daffodils, fritillaria, alliums, or glory-of-the-snow (*Chionodoxa*) that are unpalatable to your wild critters. In addition, if you select varieties that bloom early, in mid-season, and in late season, you can have a spring bulb display that lasts for more than two months!

Grasses

Grasses fall into several categories: they are either running or clumping species, and they are described as either warm or cold season grasses. If you want a garden that is easy to maintain, select *only* clumping grasses. These will grow larger in place rather than "running" throughout your garden bed. There are some stunningly beautiful grasses that are, unfortunately, of the running sort. Although in particular instances, for example, when managing soil erosion in an uncultivated area, this might be a positive attribute, they will be a huge maintenance burden in a groomed garden bed. Select with care, and ask your nursery owner for more information if you are unsure of the habits of a grass you are considering for your garden.

Cold or warm season refers to the temperatures at which grasses develop. A warm-season grass begins to grow when temperatures exceed seventy degrees. These grasses are cold-sensitive and will collapse during the first frost in the fall. Since many *coastal* cold season gardeners rarely experience summer

◂ Tulips are truly exceptional and offer a mind-boggling range of choices. Some determined gardeners have been successful at thwarting burrowing animal damage by planting tulip bulbs within a wire enclosure.

▾ With careful planning, you can enjoy daffodils for an extended period, from early spring until well into summer.

▲ The ethereal inflorescence of grasses adds motion and grace to a garden.

temperatures above seventy degrees, these beauties will be disappointing for them. If you enjoy hot summers and mild, protracted falls where you garden, you will be able to revel in the fabulous displays of the warm-season varieties. I envy you! Cold season grasses, by contrast, begin to grow as the snow melts, reach their peak in late summer, and can stand throughout the winter unless mashed down by a wet, heavy snow. They are the best choices for gardeners who have cool summers, short fall seasons, and long winters.

Herbaceous Perennials

Plants in this category die back during the dormant season. These are the plants more commonly and simply called "perennials." Well-known examples include irises, campanulas, day lilies, monkshoods, daisies, and poppies. They have soft rather than woody stems. Your main criteria will be those already discussed in "Selecting the Best of the Best": overall health, new growth, proper hydration, and an absence of pests, be they weeds or insects.

What is important to recognize about herbaceous perennials when buying them is that they have different growth and bloom schedules. Some hold up their flowers proudly as the snow melts away around them, whereas others

▲ Plants that are blooming when your nursery first opens for the season will likely flower early in your garden in subsequent years. *Iberis sempervirens* 'Alexander's White' at Alaska Hardy Gardens, Homer, Alaska.

▲ Well-known garden plants often have intriguing cultivars. A good example is silvery-blue *Campanula persicifolia* 'La Belle', a lovely mid-summer bloomer. Kathy and Mike Pate's garden.

have barely gotten around to blooming by the time snow is about to return in the fall. Many are at their best between these two extremes. If you want the maximum gratification possible from your garden, you will need to include plants that provide beauty throughout your growing season. Early bloomers are easy to recognize; they're blooming or in bud in the nursery in spring. Many mid-season performers are household names or will be fairly well advanced by the time the nursery moves its stock outdoors. To keep your garden looking great through fall, however, you will need to recognize the attributes of "late bloomers" in the nursery.

In "Giving Late Bloomers a Second Look" you can read more about later-developing plants as well as how to evaluate them on the nursery bench. Please take the time to learn about these jewels so that you can benefit from the staggered bloom times of the huge variety of herbaceous perennials that will thrive in cold climates.

Shrubs

Shrubs should be symmetrical and have a pleasing shape. Look for an appealing branching pattern, one in which the stems grow predominantly upward or outward rather than in toward the middle. When branches grow inward, they tend to create crossing patterns with other branches. As the

▲ Though they often bloom early in the warmth of a nursery greenhouse, hardy violas will bloom their hearts out in cool summer climates even after a killing frost. The Pritikin Family's garden.

▲ Yarrow yearns to be in the soil and so will often look raggedy in its nursery pot, but it will add reliable late color to your garden.

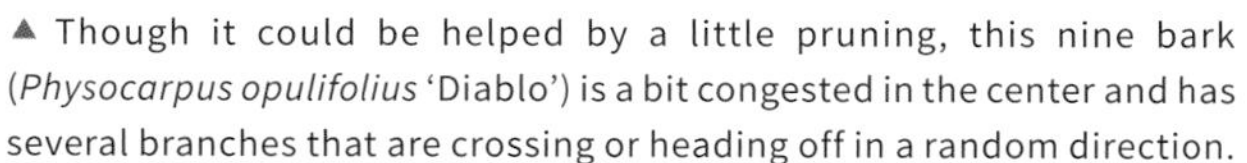

▲ Though it could be helped by a little pruning, this nine bark (*Physocarpus opulifolius* 'Diablo') is a bit congested in the center and has several branches that are crossing or heading off in a random direction.

▲ This nine bark is airier in the center, has a nice vase shape, and is more symmetrical, making it a better choice. Alaska Hardy Gardens.

shrub grows or is buffeted by wind and exposed to other weather conditions, these crossed branches will rub against each other, creating bark damage. Damaged bark on shrubs is an opening for disease and insect access.

Most shrubs have multiple stems growing from the crown. If you are considering one of these, select a well-formed shrub that has more stems without being too overcrowded. In a few cases, the branches will grow from a more central stem, branching out fairly close to the ground; in this case, look for one with a larger caliper (diameter) in this central stem. Examine the bark of the shrub you are considering to ensure that it is undamaged.

Trees

Trees represent a more substantial investment than do herbaceous perennials or annuals. Ideally, they will also be a mainstay in your garden. Therefore, they deserve an exceptionally critical examination before purchase. Begin your scrutiny by looking at the shape and structure of the tree. Separate the

▲ Rather than having a strong central leader, this selection has two competing branches at the top.

▲ This conifer has good symmetry, is well branched, and has a clear central leader. Alaska Hardy Gardens.

ones you are considering from the others so that you can carefully study them. You want a tree with well-spaced branches, not one with branches all emanating from the same location on the trunk. Look for a symmetrical shape with a good balance of branches all around the trunk. Bypass trees with a forked main trunk; these may split with maturity or the weight of snow.

In almost all cases, a strong central leader is essential. The few exceptions are trees that have multiple trunks, such as river birch (*Betula nigra.*) The leader is the uppermost portion of a major structural limb. Ideally, the highest and stoutest leader is the center one—hence it's called the central leader. It is the primary growth point for a tree and "leads" the tree upward.

When you have focused in on a particular specimen, carefully examine its bark. Avoid trees that have damaged, abraded, or cut bark. These are openings for disease and insects. Speaking of insects, examine the tree for pest insects, too. Because most nurseries display their trees outdoors, pests are not as common as they are on plants inside the warm, moist atmosphere of a

greenhouse. Nonetheless, carefully inspect a tree you are considering. Look for aphids or caterpillars. Walk away from a tree that has either.

Trees are often priced by the caliper of their trunks, although they are sometimes priced by the size of the pot. If your nursery uses the latter method, choose the tree with the largest caliper, all other things being equal.

◀ Be sure to check the ultimate size of vines, and plan for it. Even this huge *Clematis tangutica* fit into a one-gallon pot when it came into my life.

Vines

When selecting a vine, carefully make note of its expected size. Some very unassuming-looking vines will grow at an incredible rate once planted and overwhelm the trellis or other structure that you might have planned for their support. Be sure the plant will fit comfortably into your space, or you will find yourself spending a lot of time pruning. When examining a woody perennial vine at the nursery, look for multiple stout stems at the base of the plant and new growth along the stems. As with anything else, look for all the signs of health and vigor.

GIVING LATE BLOOMERS A SECOND LOOK

In the introduction, I mentioned the easy seduction of plants in bloom and the resulting temptation to buy them. I also noted that some perennials, especially herbaceous ones, develop at different times during the season, making some stand out more at the nursery. Naturally, the earlier bloomers will look better, fuller, and more like the mature plants they will ultimately become.

Late bloomers, on the other hand, not only bloom after others have completed their showy stage but also often emerge from the ground later than other herbaceous perennials. This makes them excellent partners for bulbs, because their tardiness in breaking dormancy allows bulbs to strut their stuff early in the season. Late bloomers appear just in time to conceal the tattered foliage of those early bulbs, which should be allowed to die back naturally. Unfortunately, late bloomers are also often late to develop in the nursery. This is especially true in remote locations like Alaska, where, to save on shipping costs, many nurseries buy plants as "plugs" or "bare roots" rather than as mature plants in large pots filled with heavy, expensive-to-ship soil.

This presents a dilemma for both nursery owners and shoppers alike. If we buy only plants showing their buds or flowers, we will have a wonderfully colorful spring garden, but little else as the season progresses. Nursery owners who offer a full range of perennials are often left with inventory that didn't have shelf appeal during the crazed shopping days of spring. Gardeners who learn to recognize these late bloomers and their value will be rewarded with a garden that is fabulous throughout the season, from spring all the way to the first hard freeze of early winter.

Regrettably, few visual clues are available to help you determine whether a particular plant is simply not doing well and languishing or is perfectly healthy but a late developer. The most illuminating feature is not an obvious one on a potted plant; it is the root structure. A sickly plant will likely have little root growth, whereas a healthy late bloomer will have a viable-looking,

◀ Not only is the color of gentian flowers an amazing blue, aptly called gentian blue, but it also shows up at the very end of the garden season, as you can see by the fading foliage of the surrounding plants.

▲ Late-blooming plants can often look poorly in the nursery. Though the one in the back left of this photo would be my first choice, all should prove successful in the garden. If you look carefully at the front left one, you'll see at least eight new shoots coming up. Alaska Hardy Gardens.

▲ Not all plants that develop late at the nursery are late bloomers. Many northern nurseries pot up bare-root plants and tubers, as is the case with these peonies, which will bloom mid-summer. In this instance, look for the one having the most "eyes" (growth points). Alaska Hardy Gardens.

▼ Bugbane (*Actaea simplex* 'Hillside Black Beauty') often looks sad at the nursery in spring, but it has fabulous foliage all season in the garden and blooms after nearly everything else is finished.

actively growing root ball. Not all nurseries will appreciate your tilting the pot over to slide the soil and root out for an inspection, but doing so will indeed reveal a solid clue. A preferable and more reliable way to discern one from the other will necessitate some diligence and study on your part. If you learn which plants develop and bloom later, you can go right to them at the nursery. Knowledge will be your guide. To help you get started, I've included some very special late-blooming plants in the next section. Look for and read about them so that you'll have the expertise to spot these end of season stars on your next nursery visit.

Beyond reading this book, there are additional things you can do to increase your knowledge about great plants. Go on garden tours throughout the season, and ask about the plants that attract you. Make notes about these plants, including the time of year when you see them. Botanical gardens can also be a great resource in your quest for knowledge. Visit them frequently. When I decided to start gardening in Alaska, I had to learn an entirely new set of plants from those I had used in my gardens in southern California and Arizona. One way I did this was to visit the display gardens at my local nursery at least every two weeks, taking photos and making notes about what I saw there that appealed to me. I also volunteered there, helping with everything from filing plant tags to potting up new plants for sale. All this takes time to learn, but it's an enjoyable quest. Keep in mind that creating a beautiful garden is an ongoing process. As you learn more, you can always make changes that will enhance your results and make your garden a joy for you and the envy of your friends—all season long.

What are we waiting for, then? Let's turn to the next section and start learning about the coolest cold-climate plants!

SECTION V

Cool Plants for Cold Climates

In this section, plants are grouped in familiar categories: annuals, bulbs, grasses, herbaceous perennials, shrubs, trees, and vines. Within each, you'll find plants listed by botanical name. *Oh, no,* you might say, *not all that Latin!* Yes, sad to say, but true, and here's why: Even though common names are often more fun, colorful, and descriptive and are certainly easier to pronounce, they can lead to confusion and disappointment. Often there are many common names for the same plant, causing confusion. Worse, the same common name is often used for more than one plant, creating potential disappointment. In order to identify plants accurately and uniquely so that you find and receive exactly what you want, especially if ordering online, use the scientific or botanical name. In this system, a unique name refers to one particular kind of plant and to only that plant, no matter where in the world.

The makeup of a scientific or botanical name always includes its genus and species. You might see this referred to as binomial (two-name) nomenclature. You can blame this whole thing on a Swede by the name of Carl Linnaeus who lived long ago (1707–1778). Although those of us unschooled in Latin might struggle with scientific names, his system was a massive simplification of what was going on during a time when many new plants were being discovered. Before Linnaeus, it was pure chaos.

As more and more plants have been discovered and developed, scientific names have been extended. Sometimes you'll also see a variety or a cultivar. What *are* all these things? Let's keep this as

◄ Pretty pink nodding onion (*Allium cernuum*).

Complicated!

Whenever we make a categorical statement, something happens to present an exception. Often it is advancement in knowledge and understanding; sometimes it is merely a change in convention. These things have happened in horticulture as well.

As taxonomists learn more about the DNA of plants, they will sometimes move a given plant from one genus to another or from one species to another. The result is a name change! Usually there is a time lapse as nurseries and gardeners adopt the new designation. In such cases, I've used the name you'll most likely see at a nursery or online but have also included the older or newer name as a synonym.

It is also the case that some naming conventions, punctuation, and the like have also changed over time. For example, in the past, species were capitalized if based on a proper noun. You might see this form in older reference books, but more modern books will never capitalize the species.

So sorry—I do wish this were more straightforward. Sheesh!

simple as possible: A genus is a grouping of plants having similar characteristics. A species is a smaller grouping within the genus. In other words, a refinement of the grouping such that the members of the smaller group have more in common with each other than they do with other species in the genus. In a botanical name, the first word is the genus. It is italicized and capitalized. The second word is the species name. It is also italicized but is in lowercase. *Iris setosa,* then, is a plant from the genus *Iris* and the species *setosa.*

A variety is a naturally occurring subset of a species that is slightly different from other members *and* whose characteristics reproduce in future generations via seed. This designation is also used when a group of plants is identical to the genus and species but has a notable difference, such as foliage color. In the case of *Berberis thunbergii* var. *atropurpurea*, the variety refers to a group of barberries of the species *thunbergii* that have dark purple foliage—var. *atropurpurea*. Note that the abbreviation "var." is used rather than the full word variety.

In this book, you will find at least one name with the abbreviation "ssp." within it. This indicates a subspecies and is a refinement of a species. This often happens in geographically isolated areas with indigenous plants. The differences between subspecies and variety are subtle, often argued about, and much more important to taxonomists than to gardeners!

Cultivar is short for *cultivated variety.* They are the result of human intervention but can be random mutations created by nature and then observed, selected, and propagated by humans. Though typically reproduced in a commercial environment by vegetative propagation (be that division, tissue culture, or cuttings), cultivars can sometimes be reproduced by seed, too.

The terms *cultivar* and *variety* are sometimes used loosely and interchangeably, but the naming convention is quite different. A cultivar name follows the species, is in roman type, is capitalized, and is enclosed in single quotation marks: for example, *Echinops bannaticus* 'Blue Glitter'.

Finally, as if you hadn't had enough already, here's one more item you should know, because you will see it frequently in the names of the plants in this book. An "x" within the botanical name of a plant indicates a hybrid or a "cross" in which the resultant plant has some of the characteristics of each parent. To achieve the same result, the offspring must be reproduced vegetatively.

So then why have I been dragging you through all this technical information? First, I want the names you see in this book to make sense. Also, when I refer to species, genus, or cultivar in a description, you'll know what that means. For example, when I say that a "cultivar" has some specific improved characteristic over the "species," I'll mean that the selected or created, cultivated version of a genus and species is different from, and better in some way than, the naturally occurring genus and species. But enough of this!

The plants described here are all ones I view as exceptional or "cool." In some cases, I've written about only one cultivar of a genus and species, because it is the standout among its close relatives. In other cases, you'll find descriptions of several members of a genus or of a species. In these cases, there is a brief description of characteristics of the genus that apply to all the selections to follow. Then, in the description of each specific plant, the special qualities of that plant are highlighted.

In each plant description, you will find one or more common names below the botanical name. Sometimes there are more of these than makes sense to list, so you might know a plant by a different common name than you'll find listed. You can find plants in the index by either common name or botanical name.

ANNUALS

Here are just a few unusual "annual" suggestions to encourage you to try something very different in your containers and beds. Most of these plants are sold and treated as annuals in cold climates because they will not survive the winter outdoors, but they are actually perennial in their native habitat. This gives you the opportunity to overwinter them indoors or to take cuttings for the next year. Alternatively, toss them onto your compost pile at the end of the season, and start with a fresh slate the next spring. The point is, you're going to live with these plants for only a year, so try something new—perhaps even outside your comfort zone—and enjoy the results!

◀ Mixing colorful annuals such as this bright orange calendula with herbs and other edibles will add an element of surprise to your combinations.

▲ Wonderfully textural straw flowers (*Bracteantha bracteata* 'Mohave Deep Rose') are set off nicely by the dark foliage of an amaranth.

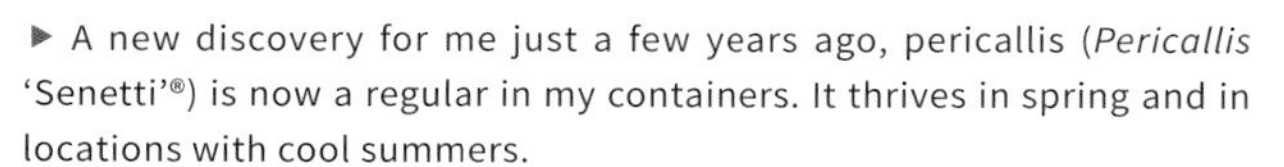

▶ A new discovery for me just a few years ago, pericallis (*Pericallis* 'Senetti'®) is now a regular in my containers. It thrives in spring and in locations with cool summers.

▲ 'Chinese Giant Orange' amaranth is an orange-lover's dream with its burnt orange flowers, bright orange stems, and apricot foliage.

▲ *Amaranthus caudatus* 'Love-lies-bleeding' is fun in the garden and is great in the vase as well.

Amaranthus hypochondriacus 'Chinese Giant Orange'
'Chinese Giant Orange' Amaranth

- Full sun
- Moist, well-drained soil
- Blooms all season
- Height 2½'–8', width 12"–24"
- Annual

Though listed as a giant that grows to eight feet, my plants average only thirty inches in a large container during the cool summer temperatures of coastal Alaska. I've seen photos of them towering over folks, so those of you with hot summers might need to give this amaranth a special place, like that of giant sunflowers. Then stand back! Though these plants may be grown purely for ornament, the leaves can be cooked and eaten like spinach. The seeds are also edible and are often made into porridge, but I prefer to leave them on the plant as nourishment for the birds.

Nearly everything about Chinese orange amaranth is one golden-orange shade or another. The stems are orange, the leaves green to golden, and the flowers coppery-orange. It's a lovely melding of colors, each of which has a gauzy softness about it. Flowers grow between the leaves up the stalk and fill out more boldly toward the top of the stem. They have an abundant, but somewhat topsy-turvy, look. It's simply a fun plant and is not one that you'll see in everyone else's garden.

There are a host of other amaranths to try. One of my favorites, *Amaranthus caudatus* 'Love-lies-bleeding', is festooned with long, draping deep red flowers that look ever so much like dreadlocks. They are nicely complemented by light-green leaves, though in cool weather these also turn red. This plant was very popular in Victorian-era gardens and is fabulous in bouquets.

Leaves are edible as boiled greens or as a nice flavor in soups. The seeds can be popped like popcorn or ground into flour. They are an important food source in South America, particularly in the Andes, as well as in India. So grow them for their beauty all summer, and then try a taste at the end of the season. (Remember, this section is here to encourage you to experiment!)

Bidens ferulifolia
Bidens, Tickseed

- Full sun
- Moist, well-drained fertile soil
- Blooms all season
- Height 12–15", width 20–24"
- Tender perennial (Zones 9–11)

A member of the sunflower family, bidens will add a bright splash of lemon-yellow to your combinations. Small, star-shaped flowers having anywhere from five to eight petals will bloom profusely and continuously without deadheading from planting until frost. If you prefer a very tidy look, snip off the spent flowers. The medium-green foliage is ferny and soft. This plant is the very definition of a filler and a spiller. As such, it makes a fabulous container plant, with its pretty flowers weaving their way among the other plants in the mix, creating vibrant combinations of color. They are particularly gorgeous with dark purple petunias. Bidens is heat- and drought-tolerant. It is also lightly fragrant and is very attractive to bees.

▲ Long-blooming even without deadheading, easy-care bidens weaves its way among the other plants sharing its container.

Cerinthe major 'Purpurascens'
(synonym *Cerinthe major* var. *purpurascens)*
Blue Honeywort, Blue Kiwi

- Full sun
- Average to low water needs
- Blooms all season
- Height 18–30", width 15"
- Half-hardy perennial (Zones 8–9)

Part of the fun of annuals is that we can try something new each year—new plants, new colors, new combinations. If you are of that frame of mind, then this is definitely one to try. Actually, *Cerinthe major* is not new at all. It was grown in sixteenth-century England, where it was valued for the purpose of sipping its nectar—by people! After centuries of absence from gardens and the development of the much more colorful cultivar 'Purpurascens',

▲ The incredible coloration of the bracts and flowers of blue honeywort will be a showstopper in your garden. Denice and Roger Clyne's garden.

▲ Two of my favorites in a combination created by Denice Clyne.

this interesting plant is making a delightful return to popularity, this time for its unusual beauty.

Rounded blue-green leaves, marbled with gray, whirl around upright stems that are eighteen to thirty inches tall depending on growing conditions. It will be larger if grown in the ground but a bit smaller if container-grown. As attractive as the foliage is, the real treat is found at the tips of the branches, where small, deep purple bells hang down, framed by saturated lavender-blue bracts that seem to have an iridescent inner glow. The coloration, which deepens with cool nights and gets better and better as the season continues, is like nothing else in the garden. It is absolutely stunning. Plant it in a large container for up-close impact. It will receive rave reviews.

Blue honeywort is nectar-rich and is very enticing to bees and hummingbirds. As frost approaches, cut off some of the branches and put them into a paper bag and store it in a dry place. The large black seeds will mature and be ready for you to propagate for another season. Alternatively, you can try letting the plant self-sow.

Osteospermum
African daisy, Cape daisy

- Full sun
- Moist, well-drained soil
- Height 1′, width 12–15″
- Blooms all season with deadheading
- Tender perennial (Zones 10–11)

▼ Dead-heading will increase the abundance of blossoms, each of which will last for a long time.

► Tangerine with pale pink centers, this *Osteospermum* looks great with purple, rose, cream, or darker orange.

►► One of the things that make African daisies more exciting than traditional daisies is the contrasting, often brilliant color of their centers.

For a refreshing new take on traditional daisies, try *Osteospermum*, or African daisies. Though the flowers are arranged in the same basic structure as other daisies, with a central disc encircled by ray petals, they are available in an extensive range of both pastel and saturated colors. The disc is often purple

in the center and adorned with bright yellow around the edges. Some discs repeat the color of the petals; others complement them with a completely different tone. Petals are available in white, cream, bright or soft yellow, and a broad range of pinks, lavenders, and purples. Apricot, orange, fuchsia, burgundy, even a few in blue, as well as many two-tone options, can also be found. As if this didn't give you enough color choice, the undersides of the flower petals are also colorful and can be quite different from the top side. Because the flowers of Cape daisies close on cloudy or rainy days, this underside hue can play a subtle role in your garden and container designs.

Rhodochiton atrosanguineum

Alaska State Fair Vine, Purple Bell Vine

- Part sun to full shade
- Moist, well-drained soil
- Height 10′, width 1′
- Blooms all season
- Tender perennial (Zones 9–11)

▲ Gather the large round seeds that form in many of the "bells" in fall and start them in a warm location under lights no later than December for a full plant the following spring.

For many years, fuchsias were my plants of choice for hanging baskets even though they required incessant deadheading to keep them in bloom. It was a time-consuming and tedious task. Then I discovered *Rhodochiton*! This amazingly versatile vine blooms all season without deadheading, requiring minimal care beyond watering. A lavish profusion of pretty bell-shaped burgundy flowers adorns the length of the trailing and vining stems. Some of the flowers produce an elongated darker center. Think male and female flowers.

Sweet heart-shaped leaves emerge green. Each leaf is borne on a twining stem that will scramble up a trellis or over a nearby shrub. Alternatively, just let this Mexican native spill from a hanging basket, where it will sway rhythmically in a gentle breeze. I've found that they can also be attractive in a large or elevated pot although, in this case, you might have to cut it back from time to time, because it can grow to ten feet.

In cool weather, the foliage takes on a burgundy flush. At season's end, you'll find firm marble-sized seedpods hidden in the female flowers. If you enjoy starting plants from seed, let these ripen, then start them under lights in December for a great show the following season. Since this vine is a tender perennial, some folks have also successfully wintered over the vines from year to year, but I have not done so myself.

▲ *Rhodochiton atrosanguineum* is a fabulous vine that performs all season. Just add water!

BULBS

The term *bulb* is commonly used to describe a group of plants that have different kinds of fleshy underground structures, including true bulbs, corms, tubers, and rhizomes. Although it's not terribly important that you know which is which, here are a few examples to give you a feel for the differences: tulips and daffodils are true bulbs; crocuses are corms; begonias are tubers; trilliums are rhizomes. While their structures vary in shape and makeup, they share a common trait: all store the substantial carbohydrates necessary to produce early and rapid growth. This section focuses on special fall-planted, spring-blooming bulbs rather than those fleshy-rooted plants that are typically potted up and sold by nurseries as perennials, such as irises, lilies, and peonies.

In cold climates, hardy bulbs do exceptionally well in the cool and sometimes downright cold air and even in the frosty, thawing soil of springtime. In northern and high-altitude gardens, they are an easy and enjoyable way to get the garden off to an early start, sometimes before the snow has completely melted. Plant them in well-drained soil and in a location where you will not be tempted to cut back their foliage before it yellows and dies down naturally. Leaving the foliage intact is essential to the replenishment of the bulb's carbohydrates, which are necessary for next season's show. Here are a few stars among the thousands you can try.

◄ Tulips are so beautiful that it's worth trying to grow some even though many critters will try to have them for dessert. This beauty is *Tulipa humilis* 'Persian Pearl'.

▲ An incredibly fragrant daffodil, 'Manly' blooms in late spring. Zones 3–8.

◄ What a welcome treat to be greeted by a river of crocus in full bloom as the snow melts away.

▲ *Allium aflatunense* 'Purple Sensation' attracts many kinds of bees.

Allium aflatunense 'Purple Sensation'
Ornamental Onion

- Full sun
- Well-drained acidic or alkaline soil
- Early summer
- Height 36", width 6"
- Zones 4–6

Though often grown as annual bulbs in very cold climates, this cultivar, like many other ornamental onions, is fun and popular. Hundreds of tiny, star-shaped light purple flowers are arranged in round baseball-sized umbels. Spherical shapes in a garden are rare and consequently provide extraordinary value from a design perspective; these are emblematic of that fact. They contribute greatly, especially en masse amid a perennial bed. The flower orb is held aloft a good three feet on a stout green stem. Substantial basal foliage emerges first but plays only a minor supporting role to the dramatic, eye-catching flowers that deservedly steal the show.

Chionodoxa forbesii
(synonym *Chionodoxa siehei* and *Chionodoxa luciliae*)
Glory-of-the-Snow

- Full sun to part shade
- Well-drained, fertile soil
- Early spring
- Height 3–7", width 2"
- Zones 3–8

The common name of this sweet little bulb derives from the fact that it is found among snow patches in the mountains of Turkey. It often blooms in cold-climate gardens as the snow is still melting away and, true to its name, brings welcome glory to the waning snow. Each bulb puts up a stem with a spray of five or more star-shaped flowers amid narrow upright to slightly arching green leaves. *Chionodoxa forbesii* is a very clear and warm blue, with a white center. 'Blue Giant' is a slightly larger and taller version, while 'Pink Giant' is also larger but has delicate pink and white flowers. All are quite vigorous and will naturalize if placed in a location that suits their needs. The

▲ *Chionodoxa forbesii* 'Blue Giant' is noticeably larger than the species and is a slightly paler blue.

◀ Glory-of-the-snow comes up through a sedum whose growth will quickly hide the fading foliage of this pretty and very early bulb.

foliage gradually dies down and all but disappears after blooms fade. Glory-of-the-snow does not do well in hot, dry summers.

Fritillaria meleagris

Checkered Lily, Guinea Hen Flower, Snake's Head

- Full sun to part shade
- Humus-rich, well-drained soil
- Late spring
- Height 12–15", width 1"
- Zones 3–8

▲ *Fritillaria meleagris* in a mix with a solitary white 'Aphrodite'.

I can't say I'm a big fan of the third common name for this plant. What I can say, though, is that I am a huge fan of checkered lilies (happily, most folks call them by this more descriptive name). Pendant, bell-shaped flowers are a little more than an inch long and half as wide. What makes them distinct is the unusual burgundy and off-white checkered pattern that covers each bell. Typically, one flower per stem dangles from the tip of a tall, very slender but rigid, caramel-colored stem with a few equally narrow, curved gray-green leaves near its top. It will naturalize in your garden or in your lawn. Though it does best in humus-rich soil, it can be successfully grown in rock gardens or between shrubs. *Fritillaria meleagris* is sometimes sold in a mix that includes the classic checkered look, a pure white form named 'Aphrodite', and a solid burgundy one named 'Charon'.

Muscari

Grape Hyacinth

Muscari produce interesting, long-lasting flower heads about the size of a grape and made up of individual flowers that, in turn, may be round, tubular, or bell-shaped, often with a constricted mouth. These spring bulbs produce flowers mostly in shades of blue with a few paler options, including one that is pure white. They are fall-planted and spring-blooming and are extremely easy to maintain, often naturalizing in the garden. All are resistant to deer, moose, and rodents.

▲ *Muscari armeniacum.*

Muscari armeniacum
Armenian Grape Hyacinth

- Full sun to part shade
- Best in average, moist, well-drained soil, but will tolerate clay
- Early spring
- Height 4-6", width 4-6"
- Zones 3-7

An heirloom from 1877, this species is wonderful for creating rivers of color in the garden, both because it naturalizes well by division and because it is very inexpensive. Clusters of cobalt blue flowers sit atop short, upright stems, surrounded by shorter, semi-upright, narrow, lance-shaped green foliage. Each individual floret is nearly spherical and has a constricted mouth edged with a thin band of white. Armenian grape hyacinths require virtually no care; they quietly bow out after blooming finishes and until next spring's curtain call.

▲ *Muscari latifolium* has a much bolder look. Plant it a bit farther into the garden so that later-developing plants will cover its fading leaves. Roni and Bill Overway's garden, Roni's design.

Muscari latifolium
Broad-leaved Grape Hyacinth

- Full sun
- Well-drained soil
- Spring
- Height 8", width 1"
- Zones 3–7

Muscari latifolium is unique in two ways. First, it has a single broad leaf rather than a group of narrow ones. Second, the densely packed flowers are arranged with dark purple on the lower half and pale blues above. Each little bloom is oblong and has a constricted mouth that is slightly darker than the rest of the flower. The strongly two-toned nature of these is simply breathtaking en masse. One of my gardening friends spent a full day at Keukenhof Gardens in Amsterdam a couple of years ago and came back raving about the incredible beauty of *Muscari latifolium*. Not surprisingly, she now has several clusters in her garden as well.

Muscari neglectum 'Valerie Finnis'
Valerie Finnis Grape Hyacinth

- Full sun
- Well-drained soil
- Spring
- Height 6–12", width 1–2"
- Zones 3–7

Named for a famous British gardener, horticulturist, and style icon in whose garden this variety was found, this is my favorite *Muscari*. Its pale blue flower racemes glow in the garden. The flower clusters are a bit longer from top to bottom than most, but that isn't the main attraction: it's simply the amazing color. As with most of this genus, the foliage is narrow and upright and surrounds the flower stems, demurely halting its growth just below the flowers, letting them serve as the main attraction.

▲ Early spring reveals *Narcissus* 'Manly' in the foreground along with *Tulipa* 'Princess Irene' in one of the author's perennial gardens.

▼ Small-cup daffodils have a completely different and more elegant look than those with the more common large trumpets. Homer Garden Club's Baycrest garden.

Narcissus
Daffodil

- Full sun
- Fertile, well-drained soil
- Spring to mid-summer
- Height 4–24", width 3–7"
- Zones vary based on division from Zones 3–7 to Zones 5–9

There are so very many options when it comes to daffodils that it's hard to choose a favorite or even several. Suffice it to say that they are a must for every spring garden in which they are hardy. Sadly, my Fairbanks friends, who commonly experience winter temperatures of forty to fifty degrees below zero Fahrenheit, tell me that they are not able to grow daffodils. For the rest of you who garden in somewhat less extreme conditions, pick the colors you love, and then plant them in the mid-point of a sunny, *well-drained* garden so that later-developing perennials will hide the bulb's declining foliage as it completes its task of replenishing the bulbs. After the bulb foliage fades to yellow, it's okay to remove it. Smaller daffodils, especially miniature ones, work best placed closer to, but still not at, the front of the border. Daffodils need no help from you other than planting them in a well-drained, sunny location and removing the flower stems after the blooms fade. Critters won't

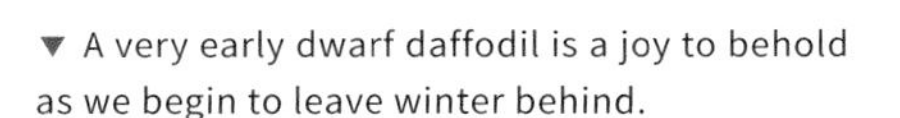

▼ A very early dwarf daffodil is a joy to behold as we begin to leave winter behind.

▼ This two-toned double is very showy and fragrant. There is an enormous number of daffodil options from which to choose!

GRASSES

The term *grass* is used, for convenience, to refer to true grasses as well as other similar plants, including carex, rushes, sedges, and cattails. What they have in common is slender, flexible foliage that moves gracefully in even the gentlest breeze, adding welcome motion to a garden. In addition, they also catch the light in an unusual way, are fabulous when backlit, and offer a quiet, whispering rustle to the sounds of a mixed garden ensemble. As a group, they are low-maintenance; just cut them back to new growth in the spring. The flowers of grasses, known as inflorescences, generally join the show late in summer, when many other plants have already peaked.

◄ Not only do these tall grasses add motion, they also create a strong sense of unity through repetition in this extensive garden at Stream Hill Park.

▼ Moisture from a recent rain sparkles on the backlit inflorescences of feather reed grass.

◄ Grasses are perfect for creating see-through effects. In this combination in Gari and Len Sisk's garden, the golden hue of the grass subtly repeats the color of the flowers of the sedum behind it.

▲ *Alopecurus pratensis* 'Aureovariegatus' growing in part shade.

▲ By the time daffodils finish blooming, spiky golden foxtail grass will be tall enough to hide the declining foliage of the bulbs.

Alopecurus pratensis 'Aureovariegatus' (synonym *Alopecurus pratensis* 'Variegatus')

Golden Foxtail Grass, Golden Meadow Foxtail

- Full sun to part shade
- Moist, well-drained, fertile, acidic to alkaline soil
- Interesting all season long, inflorescences appear mid-season
- Height 12–16", width 16–24"
- Zones 3–7

Golden foxtail grass is the perfect hardy alternative for one of the most popular shade-tolerant grasses used in more temperate climates. That popular grass, Japanese fountain grass (*Hakonechloa macra* 'Aureola'), is not hardy below Zone 5, but its golden tresses and spilling form are very desirable. No need to be disappointed though—there is an excellent alternative, adaptable *Alopecurus pratensis* 'Aureovariegatus'. It is incredibly hardy and is equally happy in part shade or sun. In shade, its variegated green and yellow leaves spill over in a fountainlike silhouette; in full sun, they are more upright and spiky. In both cases, its gleaming foliage draws attention. Flower spikes are not particularly showy and can be removed to minimize spreading and to keep the focus on the foliage.

Because of its smaller stature, this versatile grass can be used as an edging or to frame a path or entrance. It complements and coordinates well with a vast range of colors and glows when backlit. Clump-forming and slowly spreading, this cool-season grass is ideally suited to cool summer locations; it doesn't do well in hot, dry conditions. It is easily propagated by division.

Note: Although their common name is the same, this grass is *not* related to foxtail barley (*Hordeum jubatum*), which has an inflorescence that can lodge in animals' nostrils, ears, and throats, posing a danger to wildlife and pets.

Calamagrostis x *acutiflora*

Feather Reed Grass

This is a group of dramatic and upright tall, columnar cool-season grasses that require little maintenance except cutting back to six inches in early spring. Though they do best in fertile, moist, well-drained soil, they tolerate a range of soil types, including clay. Their ultimate height depends on moisture, soil fertility, and cultivar. Most will stand upright through the winter unless crushed by a wet, gloppy snowstorm. The selections presented here all have sterile flowers but are easily propagated by division. All are deer- and rabbit-resistant.

◀◀ Early in the season, 'Eldorado' is already more than two feet tall. It combines nicely with purple *Iris setosa* and *Nepeta* x *faassenii* 'Walker's Low', with blue forget-me-nots, and with chartreuse lady's mantle (*Alchemilla mollis*). Homer Public Library, Homer, Alaska.

◀ Flower spikes form by mid-season.

▼ Close up of 'Eldorado' foliage. The round, all-golden shoots are the flower stalks.

Calamagrostis x *acutiflora* 'Eldorado' PP16,486 (Patent Pending)
Golden Feather Reed Grass

- Full sun to part shade
- Moist, well-drained, fertile soil
- Attractive all season
- Height 60–72", width 15–24"
- Zones 3–9

If you like yellow, this is an excellent tall grass for you. A sport of 'Karl Foerster', 'Eldorado' has green leaves with a vivid golden stripe in the center and strong, upright honey-colored flower stems. The overall effect is more gold than green. This grass is quite arresting combined with blues or purples and is brilliant with hot colors such as scarlet or orange. Like the other cultivars of *Calamagrostis* x *acutiflora*, this one begins to grow in early spring, putting up tall flowers shortly thereafter that will stand through winter. Their bright hues make this plant a dazzling focal point above the snow.

'Eldorado' is a patented plant, meaning propagation must be licensed. Unfortunately, that makes this selection a bit harder to find sometimes, but it is definitely worth the hunt.

Calamagrostis x *acutiflora* 'Karl Foerster'
Feather Reed Grass

- Full sun to part shade
- Moist, well-drained, fertile soil
- Attractive all season
- Height 50–84", width 24–36"
- Zones 3–9

▲ As a cool season grass, 'Eldorado' becomes the star as the rest of the garden fades in late fall.

▶ Tall 'Karl Foerster' does double duty by hiding a large biofilter while adding height and motion. Gari and Len Sisk's garden.

▶▶ Tall and elegant, solid green 'Karl Foerster' creates the perfect stage for bright scarlet Maltese cross (*Lychnis chalcedonia*).

'Karl Foerster' is the best-known and tallest of the *Calamagrostis* x *acutiflora* cultivars. Originally identified in 1950 by the German horticulturalist after whom this cultivar is named, this impressive grass helped lead the popular trend of incorporating grasses into mixed borders. 'Karl Foerster' starts growing very early in spring, as do most cool-season grasses. It has stiff stems of glossy, solid green that support tall, glorious inflorescences. These sterile flowers form in early summer, beginning as mauve and maturing to a stunning burnished gold that looks fabulous with the russet and orange hues of fall. They are useful for bouquets, whether in water or a dry arrangement.

Feather reed grass is particularly effective in mass plantings, creating an effect that is riveting as a focal point. Used singly, this grass makes a dramatic exclamation point. In either case, it moves gracefully in the slightest breeze, adding delightful motion to a garden. The Perennial Plant Association honored "Karl" as its plant of the year in 2001.

Calamagrostis x *acutiflora* 'Overdam' and 'Avalanche'
Variegated Feather Reed Grass

- Full sun to part shade
- Moist, well-drained, fertile soil
- Attractive all season
- Height 50–60", width 15–24"
- Zones 3–9

More compact and a bit less columnar than 'Karl Foerster', both 'Overdam' and 'Avalanche' are excellent grasses for a mixed garden. Both have pretty green-and-white striped foliage; 'Overdam' is white on the outer edges of each blade with a green center, whereas 'Avalanche' has a single white stripe down the middle with

◀ 'Avalanche' is full and lush in early spring. Homer Veterinary Clinic's garden.

◀◀ *Calamagrostis* x *acutiflora* 'Overdam' with a pretty pink beardtongue (*Penstemon*), yellow *Primula florindae*, and a pale pink ground cover known as rock soapwort (*Saponaria ocymoides*) in Kathy and Mike Pate's garden.

green on the outer edges. Other than in their striping patterns, the two cultivars are virtually the same. The fresh new growth of their enticing leaves begins early in the spring and is followed shortly thereafter by deep mauve to pale lavender inflorescences in early summer. They are particularly attractive combined with soft pastels. As the flowers mature, they change to beautiful wheat-colored gold. The flowers will usually stand through the winter, adding scene-stealing beauty to your winter landscape. They are also excellent for bouquets and dried arrangements. In spring, hummingbirds use mature grass inflorescences in the making of their nests. 'Overdam' and 'Avalanche' are more shade-tolerant than other cultivars but do not do well in hot, humid conditions.

Deschampsia cespitosa 'Bronzeschleier'
Bronze Veil Tufted Hair Grass

- Full sun to part shade
- Tolerates soil from dry to moist, acidic to neutral
- Attractive all season
- Height 24–42", width 42–48"
- Zones 3–9

Though it's neither the showiest nor the most colorful of the hardy grasses, if I were limited to only one kind of grass forever in my gardens, I'd choose *Deschampsia cespitosa* 'Bronzeschleier' to be the one. The threadlike medium green foliage of this special grass has incredibly nice texture. It has a gracefully arching form and a prodigious amount of wispy bronze inflorescences that reach out beyond the mound of foliage. The flowers are a welcome addition to mixed cut arrangements and have a long vase life. A mature clump of bronze veil hair grass can reach four feet wide, giving this versatile grass enough presence to be used as a focal point. On the other hand, its understated beauty makes this grass an amiable companion in a myriad of situations.

The tolerant disposition of 'Bronzeschleier' extends to the range of conditions in which *Deschampsia cespitosa* can be grown successfully. It thrives in moist, well-drained or continually damp soils, whether neutral or acidic. It will tolerate a relatively hot and dry environment with ample water and performs equally well in sun or part shade. That covers a lot of territory. This is a cool-season grass that can be propagated easily by division. Cut it back in early spring, and divide it every three to four years.

▼ *Deschampsia cespitosa* 'Bronzeschleier'. Homer Public Library.

▼ The inflorescences of 'Bronzeschleier' are incredibly soft to the touch. They're wonderful in bouquets. Homer Public Library.

Helictotrichon sempervirens
Blue Oat Grass

- Full sun
- Adaptable to a wide range of soils, but good drainage is essential
- Attractive all season
- Blooms June–September
- Height 20–48", width 36"
- Zones 3–8

▲ As summer wanes, the blades of blue oat grass change to buff, blending beautifully with other fall colors. *Leucanthemum* x *superbum* 'Banana Cream' is in the background; *Actaea simplex* 'Hillside Black Beauty' (black bugbane) is in the foreground. Gail and Bob Ammerman's garden.

Though care must be taken to site this grass where drainage is impeccable, it is magnificent when grown well. It forms an attractive loose, fountain-like twenty-inch-high clump of fine blue blades of grass. Beginning in early summer, tall flower stems soar above the foliage to forty-eight inches. Soft, buff-colored tassels drape gracefully from the tips of stiff blue stems and move rhythmically in the slightest breeze. The effect is mesmerizing and lasts throughout the growing season.

The hue of blue oat grass is very similar to that of many blue spruce selections, offering an opportunity to repeat this somewhat unusual color in your garden design. It is a very easy-care plant, requiring only minor cleanup in late winter or early spring. Rather than cutting back the foliage mound, rake out older foliage with a handheld rake or, alternatively, don rubber-palmed gloves and rake your fingers through the foliage from underneath, freeing any spent blades from the prior year, and then trim back the old flower spikes. Like many blue-hued plants, this grass is drought-tolerant; it is also resistant to deer, moose, and rabbits.

◄◄ *Helictotrichon sempervirens* puts forth soft flowers in mid-summer. Kathy and Mike Pate's garden.

◄ Blue oat grass does incredibly well in warm and dry rock gardens like the one I designed for Leah Evans and Howard Cloud.

HERBACEOUS PERENNIALS

A **herbaceous perennial dies back to** the ground during winter and emerges from the ground in spring. This turns out to be an excellent survival strategy in harsh climates. These plants are usually referred to simply as "perennials."

Some herbaceous perennials will bloom the first year from seed; most will not bloom for two or three years or even longer. Perennials will come back repeatedly without replacement if they are planted in an environment that meets their cultural needs. Remember that "perennial" does not mean "eternal." Some are known as "short-lived," lasting only three or four years. Still others that have a substantial life expectancy of forty years or more can succumb to disease or decline early if their cultural needs are not met. Many herbaceous perennials require periodic dividing and other maintenance to continue to thrive. A few will bloom for several months of the season, though most present their flowers for a shorter period—three to four weeks is fairly typical in cool summers. Almost as if to make up for their shorter bloom periods, they offer a rich variety of foliage, texture, bloom type and color, size, and shape. Some hug the ground, while others have a bold silhouette, making them architecturally interesting in a garden tableau. And because they return year after year, growing lush and fuller with the passage of time, herbaceous perennial plants are often used as the mainstay of a cold-climate garden. Accordingly, this chapter has more candidates for you to consider than the other plant chapters.

▲ Peonies can live more than forty years if their cultural needs are met.

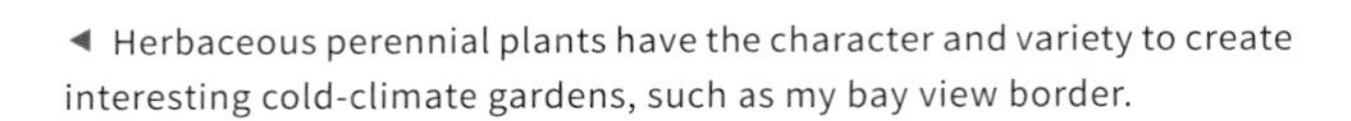

◄ Herbaceous perennial plants have the character and variety to create interesting cold-climate gardens, such as my bay view border.

► With its charming viola faces, spicy fragrance, and season-long bloom, 'Etain' is a wonderful but short-lived plant. If allowed to go to seed, it will create seedlings, although they might vary somewhat from 'Etain'.

Aconitum x *cammarum* 'Bicolor'
(synonym *Aconitum carmichaelii* 'Bicolor')
Bicolor Monkshood, Wolfsbane

- Full sun to part shade
- Humus-rich, well-drained, mildly acidic soil
- Blooms late July to early September
- Height 48–60", width 12–15"
- Zones 2–7

A very hardy, ornate, and sumptuous long-blooming hybrid perennial, 'Bicolor' can easily become the star of the show. It sports unusual-looking chartreuse buds that remind me a bit of lima beans. (As with lima beans, I will not eat them—and neither should you! The entire plant is toxic if ingested, particularly the roots.) As the buds open, they reveal striking hood-shaped white flowers with intensely blue edges. Buds and blossoms are clustered in loose panicles along the top half of stout, upright stems. Below this full array of flowers is dark green, deeply cut, almost lacy foliage, providing a rich backdrop that creates the overall look of an elegant bouquet. It's simply marvelous. From a design point of view, the mix of colors in buds and flowers gives you many choices to emphasize in your combinations, and the highly textured foliage is a great contrast for bolder-leaved plants.

▼ You will often see pollinators visiting the intricate flowers of 'Bicolor' monkshood.

▼ *Aconitum* x *cammarum* 'Bicolor' with its interesting chartreuse buds.

▼ The hood-shaped blossoms of *Aconitum lamarckii*, pictured here with *Filipendula rubra* 'Venusta', are not as large as those of 'Bicolor'.

Though it is relatively tall at four to five feet, I've not found the need to stake this plant provided it gets adequate sun. In fact, with its upright habit and narrow girth, 'Bicolor' monkshood makes a great alternative to delphiniums for those of you who abhor staking. It is decidedly easy-care in the garden, where it continues to bloom without deadheading. If its needs are met, it will thrive for years without division, expanding slowly, offering up new shoots around its circumference that you can separate in spring to share with a friend.

Other hardy monkshoods worthy of consideration include the *self-sowing* common monkshood (*Aconitum napellus*). It makes more seedlings than I like to deal with but is excellent for a wild, native plant garden. Though somewhat variable in height, flower color, and structure, it is most often a lovely deep indigo blue, approximately four feet tall and generally much bushier in appearance than *Aconitum* x *cammarum* 'Bicolor'.

A friend of mine from Fairbanks counts *Aconitum carmichaelii* 'Pink Sensation' among her favorites and says that it has been dependably hardy in her frigid part of Alaska. Another, *Aconitum lamarckii*, has pale soft yellow flowers growing to over five feet tall but needs staking or a stout neighbor on which to lean. Less commonly available is a sweet dwarf variety, a hybrid called 'Blue Lagoon'. It stands a mere twelve inches and puts forth similar purple-blue flowers starting from quite low on the stem. The result is a very showy little plant.

Allium cernuum

Nodding Onion

- Full sun to part shade
- Well-drained, average soil
- Blooms mid-summer for an extended period
- Height 18–24", width 6–12"
- Zones 3–8

For a delightful mid-summer surprise that lasts for weeks on end while requiring virtually no care, *Allium cernuum* is hard to beat. Its unassuming narrow, flat, mid-green foliage can be tucked in nearly anywhere and might be overlooked at the start of the season, but the soon-to-come flowers will strut onto the stage with real aplomb. They're arranged in sprays (called umbels) of twenty-five to forty pink flowers atop tall scapes that rise well above the foliage. These stalks are bent just below the flowers, causing them to face downward—hence the common name: nodding onions. The effect reminds me of looking into the Philadelphia sky of my childhood

▲ Nodding onions mingle amicably with sea holly, sedum, and primroses in mid-summer.

▲ *Allium cernuum*.

▶ A closer look at the fireworks-like spray of flowers at the end of each stem.

on Independence Day evening, seeing wonderful fireworks displays during which sparks of light curved downward after the initial high-altitude burst. Each flower opens at different times, creating a mix of buds and blossoms that extends the bloom period for up to six weeks.

Allium cernuum grows wild in the Ozark region of Missouri, where it inhabits rocky slopes and open woods. In a garden, these plants do best in full sun but benefit from some afternoon shade in particularly warm summer locations. Being a native plant, it may naturalize through seeds as well as bulb offsets. Although the bulbs simply increase the size of the clump, you might wish to deadhead the flowers to control any seed propagation. The foliage, flowers, and bulbs all have a mild onion scent when bruised, reminding us that these are members of the onion family. Although they are edible, I cannot imagine why you would want to eat them rather than enjoy their beauty in your garden for years to come.

Artemisia ludoviciana 'Valerie Finnis'
Western Mugwort, White Sage

- Full sun
- Well-drained, poor to moderately fertile, acidic to alkaline soil
- Foliage beautiful all season, blooms mid-summer to fall
- Height 24–36", width 24–36"
- Zones 3–9

Though 'Valerie Finnis' is incredibly beautiful and is always a huge hit when my garden is open for tours, I nearly left her out, because she's a bit of a bad girl. To her credit, she is the best-behaved of a genus and species notorious

▲ *Artemisia ludoviciana* 'Valerie Finnis' early in the season.

▲ Flower spikes accentuate the verticality of 'Valerie Finnis'.

▲ Seed heads of *Bergenia* 'Bressingham Ruby' become the stars of the early summer show with a backdrop of 'Valerie Finnis' foliage.

for rampant growth, spreading by both rhizomes and seeds. This plant is not for the timid, but if you take a firm hand in managing your garden, *Artemisia ludoviciana* 'Valerie Finnis' will wow your friends and provide you with incredible silver foliage all season long. Each pretty leaf is about three-quarters of an inch wide and three to four inches long, with five distinct points, three of which are at the terminal end. The leaves are smooth on top and slightly downy on the underside.

Though presenting a mounded form early in spring, flower spikes rise above the foliage toward mid-summer, adding a distinct spikiness to this plant's overall presence. Each flower stalk is covered with panicles of small, somewhat creamy silver blossoms. They rise substantially above the foliage and move gracefully in a breeze. Both flowers and foliage are excellent in bouquets, fresh and dried.

On its own, this selection can be an arresting focal point, but it shines in combinations. If you are creating a quiet, restful garden, combine it with pastels such as powder-blue columbine and yellow *Primula florindae*. If instead it's sizzle you seek, you'll get it by combining 'Valerie Finnis' with hot red, burgundy, or other saturated colors. In addition, you can let the spreading nature of this plant work to your advantage. If you are taming a difficult, arid slope or vegetating a large area that you'd rather not spend time managing, plant 'Valerie Finnis' with several other vigorous plants, for instance *Aconitum napellus* (described earlier), and stand back.

▲ Dramatic foliage of *Astilboides tabularis* at The Alaska Botanical Garden, a very cold Zone 3 location.

Astilboides tabularis
(synonym *Rodgersia astilboides*)
Shieldleaf

- Part shade to full shade
- Humus-rich, moist soil
- Blooms mid-summer, but the foliage is the attraction all season
- Height 36–48", width 42–48"
- Zones 3–7

This dramatic plant, once considered to be a *Rodgersia*, was moved into its own genus some time ago. Regardless of what it's called, it has star quality! Huge, round, scalloped medium-green leaves with ruffled edges are held aloft on hairy stems. Each stem attaches at the center of a leaf, forming a slight depression that will hold a dollop of moisture. As weather cools in fall, the big leaves turn yellow and then a rusty tan. Both are absolutely lovely.

In early summer, large, creamy plumes are held above the foliage on stout stems. They look like oversized astilbe plumes. Though a star, this plant is not a prima donna. It's a surprisingly easy-care plant that doesn't require division to stay healthy and productive. On the other hand, a note of caution is in order for those whose climate is fickle in the spring. As leaves begin to unfurl, they are susceptible to damage from late-spring frosts, so if you see an exceptionally cold night heading your way after the plant has started to leaf out, cover it carefully with some floating row cover or even an old sheet.

▲ *Astrantia major* 'Hadspen Blood' mixes with grasses and other perennials at the Homer Garden Club's Baycrest garden.

Astrantia major 'Hadspen Blood'
Masterwort, Hattie's Pincushion

- Full sun to part shade
- Humus-rich, moisture-retentive, mildly acidic soil
- Blooms mid-June to September
- Height 24–36", width 18–30"
- Zones 3–9

Though it's always difficult to select just one favorite plant, this would certainly be among my top contenders. The cultivars of *Astrantia major* are exceptionally attractive and versatile perennials, well adapted to cold climates. As a group, they offer one of the longest bloom times of any hardy perennial, flowering nonstop from mid-June until September without any need for deadheading. 'Hadspen Blood' flaunts masses of inch-and-a-half-wide intricate, perky, deep carmine flowers surrounded by a collar of triangular bracts. Blooms are arrayed in small clusters atop sturdy stems that stand erectly to nearly three feet tall. Foliage is fairly bold, very lush, dark green, deeply toothed,

▲ Intricate 'Hadspen Blood' flowers last for weeks when cut for a vase and hold their color well when dried.

▲ One of the white cultivars, *Astrantia major* 'Star of Billion', displays a hint of green in its flowers.

▲ *Astrantia major* 'Star of Beauty' interweaves with lavender catmint (*Nepeta* x *faasenii* 'Walker's Low') and deep purple salvia (*Salvia nemorosa* 'Caradonna') creating a soft, diaphanous combination at South Peninsula Hospital.

and palm-shaped. This exceptional foliage makes a good foil for small-leaved plants, and its deeply cut shape serves equally well as a contrast to either strappy foliage like that of the iris or bolder-leaved plants such as hosta.

Astrantia major 'Hadspen Blood' has such a strong presence in the garden that it makes a terrific focal point, yet it also blends beautifully in combination with other plants. Its versatility extends to the vase, where its cut flowers will look fresh for well over two weeks in water and maintain their color nicely in dried arrangements. Look at the flowers carefully and you'll see why one of the common names for this plant is Hattie's pincushion.

Plants gradually increase in girth, becoming as wide as they are tall, but this trouble-free beauty requires no dividing to maintain vigor. It will do well in both full sun and in part shade, although in the latter case, it will begin blooming later. Cool nights are best.

Although the species is also pretty and has blossoms of greenish white with green bracts, it self-sows and spreads too much for my liking, so I stick with the cultivars (cultivated varieties). Among these, flower colors range from deep burgundy to pink to white. Most of the accolades given 'Hadspen Blood' apply equally to other cultivars. There are many excellent dark red to maroon selections of *Astrantia major*—'Claret', 'Lars', 'Moulin Rouge', and 'Ruby Wedding'. Pink choices include 'Pink Pride', 'Florence', and 'Tickled Pink'. 'Alba', 'Shaggy', and 'White Giant' are delightful in white, some with a hint of green.

Clematis integrifolia
Shrub Clematis, Solitary Clematis

- Full sun
- Moist, well-drained soil
- Interesting throughout the growing season
- Height 24–30", width 24"
- Zones 3–9

One of my all-time favorites, this non-vining, herbaceous species of *Clematis* will both entice and confuse visitors to your garden. They'll love it but have no idea what it is—a sad circumstance for such a fabulous plant. *Clematis integrifolia* bears a single charming, bell-like flower with recurved tips at the end of each stem.

The show goes on for most of the season, beginning with a clam-shell bud that reminds me of the ravenous plant in *Little Shop of Horrors*. As the two halves of each bud begin to slowly separate, a hint of flower color is revealed, engendering curiosity until the clam shell finally pops open completely. The emerging flowers last for four to six weeks and are followed by winsome, fuzzy, white seed heads typical of many vining clematis. These persist into fall until a hard frost drops the curtain on the show.

Though there are several cultivars in varying colors, by far the best and most beautiful choice is the species. It sports blossoms of mid-blue with a touch of violet and a hint of silver. Its foliage is unremarkable, though attractive, and is covered with little hairs that give it a silvery cast. Planted in full sun in an area of cool summers, the stems need no support. In less light or warm weather, an inconspicuous support may be in order. A few of the cultivars have faded-looking flowers and tend to flop more often, so seek out the species.

▲ Because of its extended bloom period, you will often be able to enjoy buds as well as partly and fully opened flowers simultaneously.

▶ *Clematis integrifolia* in early summer.

▶▶ Fluffy seed heads follow the flowers and persist for another two months.

Dodecatheon pulchellum ssp. *alaskanum*
Shooting Stars

- Part shade in lower latitudes, full sun okay in very northern latitudes
- Moist, well-drained, humus-rich soil
- Spring bloomer
- Height 12", width 8–10"
- Zones 4–8

▲ *Dodecatheon dentatum.*

Most spring-blooming perennials present us with delicate pastel colors. What a pleasure it is, then, to find one in saturated fuchsia with a dollop of bright yellow to brighten the early garden. That touch of yellow is an added bonus in creating spring combinations with daffodils and other early yellow bloomers. The fuchsia-colored petals are swept back as if being blown by a fierce wind, while the yellow-tinged tip with its pointed stamen gives the plant its common name: shooting star. This selection will indeed be a star in your early season garden.

Foliage is substantial, oval, and rich green-blue. It forms a dense rosette from which arise chocolate stems bearing approximately six blossoms that open in series, thus providing a long period of bloom. When the flowering is complete, the foliage quietly disappears in all but the coolest locations, making way for later-developing perennials. It is an excellent choice for under deciduous trees and along the edges of a woodland garden. There are several species of this Alaska native, but their differences are so subtle that most nurseries identify them simply as *Dodecatheon* or shooting stars.

Plants will self-sow in a very well-behaved manner, creating a nice expanse of spring color. With bright, interestingly shaped blossoms, attractive foliage, salt tolerance, and an amiable habit, all in all, this will be a welcome cast member.

You may also find a sweet, delicate white variety named *Dodecatheon dentatum* (white shooting star), winner of the Royal Horticulture Society's Award of Garden Merit. Its rosette of foliage is smaller, and the stems longer, than *Dodecatheon pulchellum* ssp. *alaskanum.* With time, this sweet plant will establish a nice ground-covering clump.

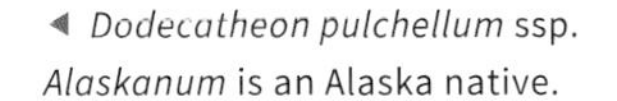

◄ *Dodecatheon pulchellum* ssp. *Alaskanum* is an Alaska native.

▲ A lovely, soft-lavender shooting star is combined with a similarly hued primrose, including matching pale yellow centers, in Susan and Don Brusehaber's Lighthouse Garden, Eagle River, Alaska. A cool combination by Susan!

▲ *Echinops ritro* 'Veitch's Blue' in a mixed border in my garden.

Echinops ritro 'Veitch's Blue'
Globe Thistle

- Full sun to part shade
- Poor, well-drained soil is best, but will perform well in most soils
- Blooms spring, summer, and fall
- Height 42", width 24"
- Zones 3–7

Spherical shapes are always eye-catching in a garden. Combine this quality with an unusual spiny look, outstanding deep blue color, and months of bloom, and you have an excellent garden plant—even a potential focal point. Each stem supports three or four blooming two-inch orbs for maximum flower power in this easy-care selection. Globe thistle is also a marvelous cut flower. If preserved in a warm environment, it will hold its color beautifully, making it very popular for dried arrangements.

Though the foliage is not dominant in the overall look of the plant, it is very interesting even so, because it, too, is spiny. The undersides are silvery and hairy, adding another dimension in a breeze. The stems are distinctly silver and very prominent, providing one more color to work with in your combinations, always a desirable attribute.

For those of you who wish to attract pollinators, this plant is bee heaven. They visit all summer but will absolutely mob the flowers late in the season.

Globe thistle does best in a warm, well-drained location and is an excellent choice for a fast-draining rock garden or steep slope. It needs little in terms of nutrients but will tolerate other soil types if you wish to add it to a mixed border.

Eryngium x *zabelii* 'Big Blue'
Sea Holly

- Full sun
- Lean, well-drained soil, tolerates drought
- All summer
- Height 36–40", width 18–24"
- Zones 2–9

Although there are many different species and even more cultivars of exotic-looking *Eryngium*, 'Big Blue' stands out among the others. Its tiny blue flowers are arranged into a large, cylindrical, cone-shaped head surrounded by huge, feathery but spiny, iridescent, sapphire-blue bracts. Bees and hummingbirds flock to these densely packed flower heads. The intense blue

▲ Sea holly should not be allowed to spill onto a path, because its flowers and foliage are quite prickly. *Eryngium* 'Sapphire Blue' is another superb cultivar and is also patented.

▲ As fall approaches, the flowers of sea holly lose the intensity of their color, eventually turning brown, but retain their attraction as focal points. South Peninsula Hospital's garden.

▲ *Eryngium* x *zabelii* 'Big Blue' takes center stage with a backdrop of *Veronicastrum sibiricum* and 'Overdam' feather reed grass in the Serenity Garden at Homer's South Peninsula Hospital.

coloration continues down the flower stalk, where even the top few leaves take on a blue tint. Below the flowers, a clump of green, spiny, basal foliage is deeply cut and has silvery lines emanating from a slightly lighter leaf center. Foliage is semi-evergreen and looks great throughout the growing season.

This fabulous plant performs in the garden for well over two months, as the flowers first appear silvery-green before donning their summer-long intense blue costumes. As the summer wanes, the flowers lose their intensity. Although the stems remain deep violet, the flowers fade to brown, surrounded by silver bracts. At any point along the way, the flowers may be cut for the vase, where they should last for weeks. They also dry well; in this case, cutting the stalks just before the flowers fully open will yield the best color.

Because this is a deeply tap-rooted plant, it does not transplant well but is quite drought-tolerant. In fact, overwatering may cause the plant to get floppy, so hold back the TLC for best results. It's great for xeric gardens (those designed to need no supplemental water) as well as for rock gardens. If you garden with a lot of climatic moisture, very careful siting for maximum drainage will help you achieve better results. *Eryngium* x *zabelii* 'Big Blue' is a patented plant whose seeds are sterile. It's a cross between *Eryngium bourgatii* and *Eryngium alpinum*. Another notable cultivar that is likely an offspring of this fabulous cross is a Blooms of Bressingham introduction called simply *Eryngium* 'Sapphire Blue', which is considered hardy to Zone 3.

Filipendula
Meadowsweet

Those who have read *There's a Moose in My Garden* will know that the genus *Filipendula* contains one of my favorite plants, the tall and elegant *Filipendula rubra* 'Venusta', the queen of the prairie. At nearly eight feet tall, and topped by huge plumes of soft pink, it definitely makes a statement in a garden. Happily, there are other amazing members of this genus whose smaller size makes them a more versatile choice.

The two others described here are easygoing plants that are long-lived and that rarely need dividing. They stand erect without staking even in breezy sites. They do, however, require a reasonable amount of moisture to do their best. All can be used as focal points or mixed with others to create remarkable combinations.

▲ The star of the meadowsweets, *Filipendula purpurea* 'Elegans', combines nicely with *Astrantia major* 'Hadspen Blood'. Pritikin Family's garden.

Filipendula purpurea 'Elegans'
Japanese Meadowsweet

- Full sun to part shade
- Moist, well-drained to somewhat boggy, humus-rich soil
- Mid-summer into fall
- Height 36–42", width 24"
- Zones 4–7

Filipendula purpurea 'Elegans' is special for many reasons. Its luminescent foliage is such a yellow green that it borders on chartreuse, but it retains a softness that is sometimes lost in full-blown chartreuse coloration. The five to seven "fingers" of its palmate shape are longer and much more exaggerated than those of others in this group. The combination of size, shape, and color of the foliage gives it true star quality.

But there's more! The stems are brilliant scarlet, standing out and contrasting wonderfully with the foliage. To literally top off this showy plant are full panicles of deep rose flowers. The overall effect is simply breathtaking. It is a gardener's dream.

Filipendula rubra 'Venusta'
Queen of the Prairie

- Full sun to part shade
- Moist, well-drained to somewhat boggy humus-rich soil
- Summer into fall
- Height 8', width 4–6'
- Zones 3–8

Queen of the prairie (*Filipendula rubra* 'Venusta') is a statuesque plant with large, showy, frothy light pink panicles appearing from mid-summer to late in the season. Its flowers will persist until after the first hard freeze, or you can cut them for drying. The leaves are nicely textured, medium-sized, and amply arranged along the stem from close to the ground to about a foot below the flowers. The absence of leaves near the top of the flower spikes adds more emphasis to the wonderfully full blooms and makes them easy to arrange in a vase. Better yet, this attention-getting plant stands tall even in windy locations without assistance of any kind.

Filipendula rubra 'Venusta' is a resilient easy-to-grow plant that can be used as a sublime focal point or as a backdrop in a large garden. In either case, give it lots of room in which to form a large clump. Full sun, a bit of humus in the soil, and consistent water are all you need to keep this stellar selection happy. My thirteen-year-old clump shows no sign of needing dividing. On the other hand, you can easily propagate *Filipendula rubra* 'Venusta' by taking plantlets from around the edges of the main clump. This selection can also be grown in boggy areas, where it will spread more rapidly.

Filipendula ulmaria 'Variegata'
European Meadowsweet

- Full sun to part shade
- Moist, well-drained to somewhat boggy humus-rich soil
- Mid-summer into fall
- Height 36–48", width 24"
- Zones 3–7

Filipendula ulmaria 'Variegata' has spectacular foliage. The leaves are of medium size and are distinctly palm-shaped. What sets them apart from others is the

▲ *Filipendula rubra* 'Venusta' enhances an attractive late-season scene in Sharon and Jerry Froeschle's garden; Sharon's design.

◄◄ *Filipendula ulmaria* 'Variegata' in the Alaska Botanical Garden in Anchorage.

◄ The non-variegated form of *Filipendula ulmaria*.

prominent but irregular creamy variegation that roughly follows the veins on each dark green leaf. A serrated edge adds even more character to the foliage.

As with other meadowsweets, the flowers are a mass of tiny individual blossoms gathered in an arrangement called a panicle. The result is a showy, soft, and frothy display held serenely above the foliage on stout stems. The flowers of this cultivar are creamy white, emphasizing and repeating the variegation in the foliage.

▲ *Geranium sanguineum* 'Vision Violet'.

▼ 'Vision Violet' makes a great ground cover around plants, such as this Asiatic lily, that might have beautiful "faces" but whose "ankles" are unattractive. Pritikin Family's garden.

Geranium

Hardy Geranium, Cranesbill

Because their common name is the same, these are sometimes confused with *Pelargoniums,* window-box geraniums. Trust me—they are totally different! That said, there are so many different hardy geraniums available that surely everyone has a favorite. To be honest, I've had several favorites over the years, each of which has eventually been replaced in my favor by another. One of these, *Geranium wallichianum* 'Rozanne', is drop-dead gorgeous and blooms incessantly all season, but, alas, its performance and hardiness proved to be uneven in Zones 3 and 4. After more than twenty years of gardening in Alaska, I think I have finally settled on two as the best of the best for very cold climates. As you'll see, they fill completely different niches in the garden.

Geranium sanguineum 'Vision Violet'

Hardy Geranium, Bloody Cranesbill

- Full sun to part shade
- Humus-rich, well-drained soil
- Summer and fall
- Height 14–18", width 24–36"
- Zones 3–8

As a dependable low-growing supporting cast member, this is a gem. It has a slightly mounded, sprawling nature with deeply divided, frilly and delicate-looking, medium green foliage. It makes a great foil for large-leaved plants and for spiky foliage as well. When the weather cools in fall, the leaves turn a rich red, sometimes even burgundy.

Throughout the summer, this pretty mat of foliage is adorned by upward-facing, cup-shaped rosy or magenta flowers. I especially appreciate it when low-lying flowers look up at us so that we can enjoy their beauty more easily. Each blossom has a small white center and deeper-colored veining, providing several choices of color to play with in combinations. These pretty five-petaled, self-cleaning flowers need no deadheading to encourage continued blooming.

Though this easy-care plant spreads via rhizomes, it does so very slowly. I've grown clumps for more than ten years without dividing them, and they're still healthy and floriferous.

Geranium x *magnificum*
Showy Geranium

- Full sun
- Average, moist, well-drained, slightly acidic soil
- Blooms all summer, good foliage color in fall
- Height 30–36", width 30–42"
- Zones 3–8

Big, bold, definitely showy, covered in blooms, and sterile! In the genus *Geranium*, being sterile is a real bonus, for many geraniums self-sow to an annoying level. Not this one.

Overall, *Geranium* x *magnificum* has a round silhouette and stays nicely upright. I've read that it flops in wind and rain in the heat of the south, but I've never experienced that in cool Alaska gardens. The foliage is bigger and bolder than that of many hardy geraniums and has a nice scalloped look, with six or seven lobes all connected at the center. Medium green throughout the summer, the leaves turn a brilliant red in crisp fall weather.

▲ *Geranium* x *magnificum*. Gari and Len Sisk's garden.

▼ A closer look reveals the interesting pattern of darker purples in the center of each blossom. Buds and flowers continue for an extended period without needing deadheading.

The flowers are numerous, always fresh-looking, and lovely. Happily, they stay nearly all summer. Each is cup-shaped, has five overlapping petals, and faces outward and slightly upward. For a person who likes purple, this is nirvana. Blossoms are intensely violet, with significant veining in darker purple, and have a tiny pale center.

Though I always enjoy the seed heads for a few weeks in the late summer, this plant looks tidier if deadheaded when the flower show is over. Rather than cutting each individual flower, however, reach down into the foliage and you'll find that several flower stems meet at a common stalk. It is noticeably thicker than the foliage stems. Cut that instead to reduce the time required for deadheading by 80 percent. One potential problem that has developed of late with this one—and with hardy geraniums in general—is that porcupines seem to have developed a taste for the foliage. In my garden, the local clan of these darling, but destructive critters starts at the north side of my home and slowly works its way around the house, devouring the foliage of another geranium every couple of days. The good news about *Geranium* x *magnificum* is that it recovers quickly, putting out fresh foliage in a matter of days.

Geum 'Totally Tangerine'
(synonym *Geum* 'Tim's Tangerine')
Avens

- Full sun to part shade
- Does well in most soil types
- All summer
- Height 30", width 18"
- Zones 4–7

This is an exciting plant, not only because it sports tangerine—not orange—flowers that even those gardeners most resistant to orange find attractive, but also because it does so with wild abandon all season long. What's more, this profusion of flowers is sterile, so there will be no unwanted seedlings. Flowers are held thirty inches above a rosette of strawberry-like dark green foliage on flexible stems. The effect is one of blossoms floating in air. It's quite delightful, especially when allowed to mingle with complementary neighbors.

The amount of blooms cannot be exaggerated. There is no other *Geum* like it! It's very amiable in terms of soil pH, fertility, and drainage, doing well in a large range of conditions. It excels in tough sites and does great in large rock gardens and mixed borders alike. I'm told it also performs much better in hot, humid conditions than others in the genus.

Long stems and staying power make this a superb cut flower. Butterflies also find it a must-visit destination. In fall, the rosette of foliage turns bright burgundy, adding a note of late-season color.

▶ Several shades blend to produce soft tangerine blossoms.

▶▶ *Geum* 'Totally Tangerine' adds season-long zest to the public garden created and cared for by the Homer Garden Club.

Hosta

Plantain Lily, Hosta

One of the most popular shade plants in the world, *Hosta* has a different role in many far northern gardens; here, it can be a full sun plant! In fact, in gardens where the soil warms very late in the season, hostas do better in sun or part shade than they do in full or deep shade. This is a pleasing turn of events for extremely far northern gardeners, because it provides us with a great foliage plant for full sun—and there just aren't enough of those. The flowers of this popular plant attract both butterflies and hummingbirds. Given enough moisture, it will do well under deciduous trees, where root competition can be fierce.

As many are well aware, *Hosta* is available in a mind-boggling array of colors, variegation patterns, and color combinations, as well as in different forms, leaf shapes and sizes, and plant shapes and sizes. In fact, for some time I toyed with not including hostas at all because they are so well known—at least in aggregate. Then I decided that amid so many options, some of the ones I find most enticing might be new to you.

Happily for those who prefer to shy away from botanical names, most plants in this genus are sold simply by their cultivar name, leaving all mention of specific epithets (species names) aside. Because they are prized for their wonderful foliage, one way to approach selecting a hosta for your garden is simply to look at the different offerings at the nursery and choose one with foliage that appeals to you. What is more difficult to discern at the nursery with a small immature plant, however, is how it will develop over time and how vigorous it will be. Be sure to read the plant tag carefully!

Hosta Resources

American Hosta Society
www.americanhostasociety.org

American Hosta Growers Association
www.hostagrowers.org

▲ *Hosta* 'Krossa Regal' has the same vase shape as *Hosta* 'Regal Splendor' but without the creamy edges.

▼ Although hostas can go for years without division, you can propagate your favorites by digging the entire clump and cutting it into pieces with a sharp spade or blade. If it is too big to lift, cut it into quarters in the ground, and then carefully lift each piece. Be sure to keep the roots moist, and water well when replanting.

▼ Hostas with thick, corrugated leaves are less susceptible to slug and snail damage. Marguerite and Flip Felton's garden.

▲ *Hosta* 'Blue Mouse Ears'. Kathy and Mike Pate's garden.

Hosta 'Blue Mouse Ears'

- Part shade to full shade in southern latitudes
- Full sun to part shade in northern latitudes
- Moist, well-drained, humus-rich soil
- Interesting throughout the growing season, blooms late summer
- Height 8", width 12"
- Zones 3–8

This miniature *Hosta* is absolutely adorable. Each tiny gray-blue leaf is slightly cupped and looks just like—yes—a mouse ear. The foliage is thick, making it more slug-resistant than most. The overall shape of the plant is round; it's a charming choice for a close-up location or among other tiny treasures. I like to include them in small rockeries tucked into a crevice or near a pretty rock.

The flowers, which open in late summer, have a unique round bud formation that swells noticeably before opening. The tubular flowers are more horizontal than many hosta blossoms, so they are more accessible to hummingbirds.

▲ 'Canadian Blue' looks great with dark-foliage plants such as *Actaea simplex* 'Hillside Black Beauty'. Kathy and Mike Pate's garden.

Hosta 'Canadian Blue'

- Part shade to full shade in southern latitudes
- Full sun to part shade in northern latitudes
- Moist, well-drained, humus-rich soil
- Interesting throughout the growing season, blooms late summer
- Height 18", width 42"
- Zones 3–8

There's something very appealing to me about frosty-blue hostas. If you also enjoy this hue, you'll find *Hosta* 'Canadian Blue' to be an excellent choice for the

▶ When this early snow melted, *Hosta* 'Canadian Blue' shrugged it off like a true Canadian and continued to put forth its flowers.

intensity of its color, especially when sited in part shade. Pretty heart-shaped leaves are deeply veined and very substantial, making this medium-sized hosta, like 'Blue Mouse Ears', resistant to slug damage. One of the things I appreciate about this selection is that it shrugs off the first several fall frosts and even an early snowfall, while many others crumble the first time the thermometer dips to freezing.

Pale lavender flowers add a welcome note of fragrance to the garden as they float elegantly above the mound of foliage. This hosta is gorgeous combined with deep burgundy or chocolate companions.

▲ 'Gold Standard' changes color during the season. It starts with more green than gold but finishes the season a brilliant gold with just a hint of green at its edges.

Hosta 'Gold Standard'

- Part shade to full shade in southern latitudes
- Full sun to part shade in northern latitudes
- Moist, well-drained, humus-rich soil
- Interesting throughout the growing season, blooms late summer
- Height 20", width 36–42"
- Zones 3–8

This classic golden hosta with its dark green edges quickly develops into a handsome clump. It works well as a specimen plant or massed as a colorful ground cover. Nicely textured foliage is slightly rounder than heart-shaped and bears a subtle sheen. Though some folks remove flower stalks from their hostas, I find the tubular flowers very attractive. The blooms of *Hosta* 'Gold Standard' are no exception; pale lavender blooms open late in the season atop impressive thirty-inch scapes.

▲ Mildly fragrant flowers bloom late in the season.

Hosta 'Patriot'

- Part shade to full shade in southern latitudes
- Full sun to part shade in northern latitudes
- Moist, well-drained, humus-rich soil
- Interesting throughout the growing season, blooms late summer
- Height 30", width 20"
- Zones 3–8

A selection of hostas would not be complete without at least one variegated in green and white. There are at least a zillion of these available, so you may have a very different favorite in this color combination. To me, the sharp difference between dark green and bold white is a perfect excuse to create a big splash in the garden. To create the maximum effect, a combination that also has bold, painterly strokes in the variegation will serve you well. This description suits *Hosta* 'Patriot' perfectly, with its fearlessly daring pattern in dark green and crisp white.

▲ *Hosta* 'Patriot'.

▲ Because *Hosta* 'Regal Splendor' holds its foliage high above the ground, it combines well with low-growing plants. Elise and Jay Boyer's garden.

Hosta 'Regal Splendor'

- Part shade to full shade in southern latitudes
- Full sun to part shade in northern latitudes
- Moist, well-drained, humus-rich soil
- Interesting throughout the growing season, blooms late summer
- Height 36", width 36"
- Zones 3–8

If I could have only one hosta, I would be sad indeed, but I would pick *Hosta* 'Regal Splendor'. The name alone is seductive, don't you think? What makes this one stand out from the crowd is the shape of the plant. Instead of a mound, it is shaped like an elegant vase. Blue-green leaves edged in gold to ivory, depending on light conditions and temperature, are long, graceful, and just a bit wavy. Because of the architectural structure of the plant, these are held high off the ground, away from slimy plant predators. The foliage forms a perfect vessel for incredibly tall sixty-inch scapes that will support pretty lavender flowers high in the air to attract pollinators. This stunning hosta will get more impressive with every passing year.

Iris
Iris

It's hard for me to envision a garden without at least one iris in it. Indeed, one iris is not nearly enough! Regardless of the species, when an iris comes into bloom, the sumptuous, often very colorful, complex flowers are simply marvelous. Fortunately, different species will bloom at different times of the growing season, so if you make your selections with care, you can have iris blooming in spring, mid-summer, and late summer while you enjoy the benefits of their wonderful foliage all season long. The strongly vertical, swordlike leaves of this garden classic cause it to stand out in any style garden. It's an excellent plant to place periodically throughout your garden to establish cohesion through the repetition of its eye-catching form.

If you are a cold-climate gardener, *Iris sibirica* (Siberian iris) will be one of your stalwarts. Its botanical name may lead you to believe that it's very hardy, and in fact it is. For over-the-top spring flower power, there's a wonderful dwarf species, *Iris pumila*. If you have a pond or a wet and boggy area, try *Iris pseudacorus*, yellow flag iris, or Alaska's native iris, *Iris setosa*, both of which will also perform well in regular garden soil as long as it doesn't dry out.

Though these might be the "big four" for cold-climate gardeners, there are so many other options that it's not possible to describe them all here.

◂ *Iris arenaria*.

▴ Stunning color and an unusual form make *Iris chrysographes* 'Black Form' a welcome addition to my garden.

For me the tall, flashy bearded iris so prevalent in the Pennsylvania gardens of my youth is totally unreliable no matter how many times I try one. I garden in Zone 3, typically with enormous amounts of snow. The moisture of all that melting snow may be more of an issue than the temperatures. Yet I've impulsively tried other irises with great success. A delightful, tiny, and very early bloomer called *Iris arenaria*, or sand iris, came my way when the legendary Alaska gardener, Mann Leiser, passed away. Another, *Iris chrysographes* 'Black Form', known simply as black iris, caught a friend's eye at a local nursery because of its incredibly dark color and velvety flower texture. Four years later, this Zone 4 iris is still with me, and I love it. I encourage you to give other species a try if one strikes your fancy, especially if you garden in Zone 4 or 5.

Iris pseudacorus 'Sun Cascade'
Sun Cascade Yellow Flag Iris

- Full sun to part shade
- Acidic soil with average moisture content to standing water
- Blooms late summer into September
- Height 36–48", width 24–30"
- Zones 3–9

This is a tall, robust plant with substantial swordlike foliage and frilly-looking soft yellow double flowers adorned with rusty brown etchings on the falls; it definitely demands notice. 'Sun Cascade' performs well in a normal garden environment but excels in bogs, at the margin of a marsh, or standing in shallow water. A large clump of yellow flag iris in a pond is breathtaking. (Unfortunately, in some areas,

▶ Pretty rust-colored etchings on their falls add charm to the frilly blossoms of 'Sun Cascade'. Gari and Len Sisk's garden.

▶▶ *Iris pseudacorus* 'Variegata'. Lorna and Curt Olson's garden.

▶ Each spring, a large planting of *Iris pumila* 'Candy Apple' adorns the entry garden at the Homer Veterinary Clinic.

▶▶ One of a myriad of colors available among the many *Iris pumila* cultivars.

Iris pseudacorus excels too well, outcompeting native plants. Be sure to check for local restrictions.) It blooms late in the season, usually unfurling from mid-August to early September. *Iris pseudacorus* is also available in a variegated form with outstanding vertically striped yellow-and-green foliage.

Iris pumila
Standard Dwarf Bearded Iris

- Full sun to part shade
- Average to dry acidic soil
- Blooms in June
- Height 8–12", width 16–24"
- Zones 4–7

The shortest, but showiest of the hardy irises is *Iris pumila*, or standard dwarf bearded iris. These spring-blooming lovelies with their large, elaborate blossoms are available in a rainbow of solid colors, such as the magnificent, rich burgundy of 'Candy Apple' and the delicate, soft peach of 'Betsy Boo'. There are also dramatic combinations, for example, 'Navy Doll' in creamy white with a splash of dark blue and touch of yellow and the bright orange with gold of 'Mauhaus'. To increase the impact of these wonderful colors, arrange *Iris pumila* in good-sized clumps. The gray-green leaves are barely eight inches tall and nearly one inch wide but are still strongly vertical. Dwarf iris is stunning in combination with spring-blooming bulbs and is always a welcome addition to the garden. Plant rhizomes in well-drained soil, so close to the surface that they are barely covered.

Iris setosa
Dwarf Arctic Iris, Wild Iris, Native Iris

- Full sun to part shade
- Tolerant of a range of soils, best in humus-rich, moist, well-drained acidic soil
- Blooms early summer
- Height 6–24", width 15–24"
- Zones 2–9

Alaska and Siberia are home to a lovely mid-sized iris, *Iris setosa*. In some parts of the world, it is commonly referred to as blue flag iris or wild iris; in Alaska, it's often called native iris. The flower color varies somewhat from blue to purple and occasionally lavender. Masses of blooms typically open in June and are followed by large, knobby seedpods that are useful in dried arrangements and that are attractive in the garden. Some folks choose, instead, to remove the seed heads; it's your choice. This local iris inhabits the shores of lakes and streams and so will acclimate to a similar environment in your garden, though it will also perform exceptionally well in regular garden soil. *Iris setosa* forms full lush clumps that will need periodic dividing, is otherwise trouble-free, and is tolerant of salt air and wind, growing wild as it does in the Aleutian Islands of Alaska.

▲ *Iris setosa*.

Iris sibirica
Siberian Iris

- Full sun to part shade
- Tolerant of a range of soils, but does best in moist, well-drained acidic soil
- Blooms in mid-summer
- Height 24–48", width 16–36"
- Zones 3–9

This species is native to moist meadows, but for far northern gardeners, it actually does better in an average garden environment than in very wet areas. Its height varies between two and four feet, depending on the cultivar. Colors include many shades of blue and purple as well as burgundy, pink, white, and yellow, sometimes in combination. There are literally hundreds of cultivars from which to choose, and in each, the stem bears several blooms that open successively. 'Silver Edge', one of my favorites, is a sturdy selection, with very large and striking blossoms in rich purple-blue and a sliver of white along the edges of each petal. 'Caesar's Brother' is one of the tallest, with exceptionally narrow, almost grasslike leaves and purple flowers held high above the foliage, whereas 'Ruffled Velvet' is deep burgundy with delightful wavy falls. Most Siberian irises bloom during mid-summer, while the taller varieties bloom a bit later.

▲ Handsome *Iris sibirica* 'Ruffled Velvet' has red-purple standards and velvety, darker purple, wavy-edged falls with a black and gold blaze. Kathy and Mike Pate's garden.

▲ *Iris sibirica* 'Silver Edge'.

▲ A pure white Siberian iris was a gift from our homesteading neighbors, the late Bob and Ann Gillas. The cultivar is unknown to me.

▶ Tall and elegant 'Caesar's Brother'.

Leucanthemum x *superbum* 'Banana Cream'
Shasta Daisy

- Full sun
- Well-drained, humus-rich soil
- Blooms most of the summer
- Height 18–21", width 24–30"
- Zones 5–9 (though I've successfully grown them in many Zone 3 and 4 gardens)

Who doesn't love daisies, at least in a vase? William Wordsworth even wrote a poem titled "To the Daisy" praising them. Reading between the lines, however, a gardener can recognize the tale of their prolific nature. In fact, the weedy and aggressively spreading oxeye daisy (*Leucanthemum vulgare*) has made some folks shy away from all daisies. That's a real shame, because Shasta daisies are well-behaved garden plants. They're the result of the hybridizing efforts of Luther Burbank, who continued to develop these wonderful plants near Mt. Shasta, California, the source of their common name, until he was seventy-six years old!

▲ Vigorous and long-lasting, the flowers of 'Banana Cream' join the production in the second half of the season. Gail and Bob Ammerman's garden.

'Banana Cream' combines the many stellar qualities of other Shasta daisies with a lovely soft yellow color, somewhat more compact growth habit, and a semi-double appearance. Its blossoms are fully four to five inches in diameter and appear to be somewhat double because of an extra ray of petals. They open pale yellow, brighten to lemon, and, with age, fade to cream. New flowers form at each axillary shoot. If you deadhead the first flush of blooms as they finish, this second round of flowers will extend the bloom time for six to eight weeks.

Though its foliage is not particularly interesting, the long bloom time, cheerful mass of blossoms, and fabulous performance as a cut flower—two weeks in a vase—make this a must-have plant. I only wish I'd left some room for a cutting garden so that I could have more of them! Clumps get larger in place and can be divided every three years or so to increase vigor.

For those of you who want a well-behaved daisy but who reside in Zone 3, try the Alaska namesake, *Leucanthemum* x *superbum* 'Alaska'. It's single, white, and a bit taller at thirty inches, but it carries all the other positive attributes of 'Banana Cream' with three- to four-inch classic daisy blossoms. Oh, and did I mention that all the Shasta daisies attract butterflies?

▲ *Lewisia cotyledon* 'Little Plum' does particularly well if tucked into crevasses in fast draining locations.

▲ *Lewisia cotyledon* 'Little Peach'.

Lewisia
Bitterroot, Lewisia

This is an incredibly special group of plants that will do best in a rock garden or other very well-drained site. The genus *Lewisia* is named for Merriweather Lewis of the famed Lewis and Clark Expedition. I was fascinated by the tale of these explorers' incredible journey through uncharted and wild territory, how they managed to travel *up*river, and the enormous number of calories they consumed daily. I am equally entranced by the genus named for Lewis.

Lewisia cotyledon
Cliff Maids

- Full sun to light shade
- Moderately fertile, sharply draining soil
- May bloom up to four times per season!
- Height 6–10", width 8–12"
- Zones 3–9

The common name of this succulent, cliff maid, is an excellent clue about the best place to site these pretty alpine plants. In nature, they are found in niches on cliff faces, which makes them a perfect choice for a rock wall or other form of rockery. In any event, place them so that their crowns will be free of moisture. Using a rock-based mulch will assist in draining moisture away from them. Obviously, water sparingly.

Thick, meaty, spoon-shaped, evergreen leaves form a tight rosette at ground level. Beginning in spring, and periodically throughout the growing season, short panicles of sweet flowers dance above the rosette. There is a nice range of cultivars in specific colors, including pink, rose, apricot, peach, plum, white, and yellow, although the species has pink candy-striped blossoms.

Lewisia tweedyi
(synonym *Lewisiopsis tweedyi*)
Tweedy's Lewisia

- Full sun
- Very fast-draining average soil
- Blooms in early spring, may rebloom in late summer
- Height 8", width 12–15"
- Zones 3–8

Among the twenty or so species of *Lewisia*, *Lewisia tweedyi* is one of the showiest. It is much larger than the more commonly available *Lewisia cotyledon*, a very pretty choice as well. *Lewisia tweedyi* has substantial shiny, dark green,

◀◀ The foliage and blossoms of *Lewisia tweedyi* are much larger than those of *Lewisia cotyledon*.

◀ When fully open, the flowers completely cover the foliage. A second bloom late in the season is not unusual though not generally quite as full as the first.

four-inch long, lance-shaped leaves that are slightly wavy along their edges. Together they form an evergreen, mounded rosette. In certain conditions, the leaves are purported to exhibit a hint of purple, but I've not observed that in cool, coastal Alaska.

Starting in spring and continuing into summer, this attractive mound is covered by two-inch-wide, delicate looking cup-shaped flowers. Each has a pale yellow center and nine or so petals that darken to apricot at the tips. The overall effect is a sweet little flower whose color I think of as peach. These blooms continue for at least six weeks. Later in the summer, a second, somewhat smaller flush of flowers appears. Fortunately, this second bloom is not dependent on deadheading. For several years I experimented by deadheading half of my plants and not doing so with the other half. Though the deadheaded plants looked a bit tidier, the number of flowers in the second bloom appeared to be the same in both cases.

Though this wonderful little jewel may be hard to find, it is definitely worth the hunt. Local rock garden societies may be a good source. Once you have one, you'll find that it can be easily divided to create more to share with your *very* close friends. *Lewisia tweedyi* has an extremely long taproot, so be careful if you dig them up to divide, and be sure to plant them in a spot with excellent drainage.

Liatris spicata 'Kobold'
Blazing Star, Gayfeather

- Full sun
- Moderately fertile, moist, well-drained soil
- Late summer into fall
- Height 24–30", width 15"
- Zones 3–9 (Some references say Zone 2)

As much as I encourage fellow gardeners to learn botanical names, sometimes the common name is so winsome and descriptive that it's hard to do so with a straight face. I adore the name gayfeather. It's another reminder of my

◀ *Liatris spicata* 'Kobold' will bloom the first season when started from a corm. If your season is short, you can start it indoors under lights, and then move it outside once the soil warms. 'Kobold' winters over reliably in Zone 3 but needs good drainage and full sun.

early childhood days tickling my dad with a real feather and giggling wildly as I darted out of his reach—but I digress.

The tall, self-supporting, lilac flower spikes of *Liatris spicata* 'Kobold' and a slightly darker cultivar named 'Floristan Violet' will make a bold architectural statement in your garden. The top twelve inches or so of each spike are densely covered with disc florets that open into fuzzy-looking short, string-like bits. The flowers are unusual in that they open from the top down, creating a long-lasting and interesting two-toned effect. Narrow, unexceptional green foliage forms a tuft at ground level and continues up stout stems to the point where the florets begin.

Liatris spikes not only make a strong vertical statement in the garden but also are fabulous in a bouquet, and they have become the darlings of the florist trade. Grow a big clump of them so that you have plenty to cut and bring indoors. These American natives are easy to care for and easy to start from corms, blooming in the first year. Regular division, every three years or so, will keep your clumps more vigorous. They attract nearly all pollinators: bees, butterflies, and hummingbirds. Seed-eating birds will seek out the seeds in the fall and winter, so leave the flower spikes standing at the end of the growing season.

Ligularia

Superb foliage is the hallmark of this genus. Plants require a moist environment and cool weather to excel and look their best. Even in cool summers, a modicum of afternoon shade will keep wilting at bay. Two species, *Ligularia dentata* and *Ligularia stenocephala*, can be used as outstanding focal points. They will also mix well with other plants, providing a great foliage contrast to smaller or frilly-leaved companions. Yet another use is as an elegant and tall backdrop for the garden.

Ligularia dentata
Bigleaf Ligularia

- Full sun or part shade
- Moderately fertile, reliably moist soil
- Foliage is interesting all season, flowers in late summer to fall
- Height 3–5', width 3–4'
- Zones 4–8 (some report success in Zone 3 if carefully sited in a microclimate)

◀◀ The fantastic foliage of 'Gregynog Gold', with its rich but understated coloration, is a welcome addition to any fall garden scene.

◀ In this combination in the Stream Hill Park entry garden, dark stems and bold leaves with incredible texture provide strong contrast but also repeat the colors of other plants for a unified look.

There are a number of cultivars of this sumptuous species, differing primarily in leaf color. Their big, beautiful leaves are round, toothed along the edges, and held aloft on dark stems. The foliage of 'Britt-Marie Crawford' is a rich chocolate on both leaf surfaces. 'Desdemona' leaves are more crinkly, greenish on top and distinctly purple on the underside on purple stems. 'Othello' also has green and purple leaves. 'Gregynog Gold', also known as *Ligularia dentata* x *hessei* 'Gregynog Gold', is not as common at nurseries but develops wonderful fall color. Its leaves are chocolate on the underside, dark green on top, deeply veined, and heavily serrated. Unless you intend to use the precise leaf color in your combinations, go with what is available at your nursery when it's time to purchase this incredibly cool plant for your garden. They're all spectacular!

With all these bigleaf ligularias, the flower blossoms are more golden orange than yellow. They are held on strong stems well above the foliage and look like a daisy having a very bad hair day. The petals seem to go every which way. Don't pass up this plant just because you don't like the flower color. Simply cut off the flower stalk when it appears late in the season. The fabulous foliage is well worth the trouble.

▼ *Ligularia stenocephala* 'The Rocket' in a border with scarlet Asiatic lilies, hosta, and lady's mantle. The neighboring and similar plant with more intricate foliage is *Ligularia przewalskii*. Though also an elegant-looking selection, it didn't make the cut as a "cool plant" because of the excessive number of seedlings it creates.

Ligularia stenocephala
Narrow-spiked Ligularia

- Part sun to full shade
- Foliage is interesting all season, flowers in mid-summer
- Moderately fertile, reliably moist soil
- Height and width vary by cultivar; see narrative for details
- Zones 4–8

The grand dame of the clan, *Ligularia stenocephala*, is commonly known simply by its cultivar name: 'The Rocket'. This elegant plant is a real star. The huge foliage is attractively heart-shaped and is deeply serrated along the edges. The stems of both foliage and flower stalks are nearly black, and oh—the flower stalks! Soaring a good two feet above the four-foot mound of foliage, numerous spires covered in sulfur-yellow flowers are truly magnificent; as a bonus, they require no staking.

Two other noteworthy cultivars were bred to bring the same excitement as 'The Rocket' to smaller gardens. Topping out at three to four feet, *Ligularia stenocephala* 'Little Rocket' is a diminutive replica of its more common relative but has even more deeply serrated foliage. The other, 'Bottle Rocket', has a little different look. Rather than being held aloft well above its foliage, the flowers begin right at the foliage. This creates the impression of a bouquet of flowers accentuated with foliage around the mouth of the vase. It's a lovely enhancement.

Meconopsis betonicifolia
Himalayan Blue Poppy

- Full sun in extremely northern gardens, part shade elsewhere
- Neutral to slightly acidic, humus-rich, moist, well-drained soil
- Late spring to summer
- Height 36–48", width 18–24"
- Zones 3–7

▲ *Meconopsis betonicifolia*, the prized Himalayan blue poppy.

What is the one plant that all gardeners who visit Alaska want to take home with them? It's the incredible Himalayan blue poppy! And who can blame them? The color of their huge, slightly crinkled and ruffled, silky flower petals is like that of the sky on a perfectly clear May day. Setting off this intense pure blue is a notable gathering of bright golden stamens. These flowers are held high on strong hairy stems that bear several blossoms as well as large, ovate (oval), hairy, gray-green leaves with toothed edges. Even the buds and seedpods are covered with a fine protective layer of rust-colored hairs.

Mature plants form vigorous, erect clumps that expand slowly with time. Although *Meconopsis betonicifolia* has a reputation for being short-lived, I'm convinced that this is because too many folks try to grow it in less than ideal conditions. Grown in environments with cool, moist summers and in humus-rich, slightly acidic soil, they can live for decades. My happy clumps, pictured in the nearby photos, are more than twenty years old, have never been divided, and are going strong. In coastal Alaska, they can be grown in full sun. If you garden where summers regularly exceed 70 degrees, you will do better growing them in part shade. On the other hand, if your summer temperatures regularly exceed 80 degrees, don't waste your money just to frustrate yourself. Save that plant money for a trip to Alaska in June so that you can see them as they should be grown!

◀◀ Not only is the color of the fully open flower of this plant incredible, but the freshly revealed blossom is also just as spectacular and much more saturated.

◀ Can you see the hint of purple and lavender in the flower color? Every once in a while, a "blue" poppy plant will bloom lavender; on even fewer occasions, it will present white blossoms.

There is no need to deadhead the flowers, since the seedpods are very charming. You may even get a few seedlings from them. Let the foliage die down naturally in fall, and remove it in spring. Add a winter mulch if snow cover is undependable. Here's the most difficult advice to follow, especially if you have no interest in deferred gratification: remove the flower stems the first year this plant is in your garden, well *before* it blooms. This will help direct the plant's energy to its roots, ultimately creating a much more robust clump that will bring you years of beautiful, amazingly blue flowers.

Nepeta x *faassenii* 'Walker's Low' (*synonym Nepeta racemosa* 'Walker's Low')

Nepeta x *faassenii* 'Six Hills Giant' (synonym *Nepeta* 'Six Hills Giant')

Catmint

- Full sun
- Dry to medium well-drained soil, even sandy infertile, neutral to alkaline soil
- Blooms June–September
- 'Walker's Low': Height 24–36", width 24–36"
- 'Six Hills Giant': Height 24–36", width 30–48"
- Zones 3–8

Although you may favor one over the other, these two distinct catmint cultivars are enough alike in the grand scheme of things that either can be used in most situations. As just noted, 'Six Hills Giant' may be somewhat wider than 'Walker's Low'. Its flower color is deemed to be a bit more intense by some, and it fares better in the hot and humid south. On the other hand, the traits they have in common are considerable and very appealing.

Billowy and relaxed but generally upright, they have square stems covered with whorls of small lavender-blue flowers from June through September. The only help they need to perform for this length of time is a good shearing in mid-season, when the first flush of blossoms starts to fade. Not only will this bring on a new rush of flowers that will last until the frosts of fall, but their aromatic gray-green foliage will also be rejuvenated. A closer look at these plants will reveal that each flower has two lips and is trumpet-shaped; the small slightly hairy leaves are intricately veined and wrinkled. Soft, amiable colors, combined with the loosely structured, diaphanous nature of these plants, make both 'Six Hills Giant' and 'Walker's Low' fabulous supporting cast members in any garden scene.

Drought-tolerant once established, they thrive in hot, dry locations, doing exceptionally well in rock gardens and other quickly draining sites. Both have sterile seeds, so they are propagated vegetatively. Though related to catnip (*Nepeta cataria*), they do not hold the same level of interest for cats—thank goodness! They do, however, attract bees, butterflies, and hummingbirds.

'Walker's Low' was named Plant of the Year by the Perennial Plant Association in 2007.

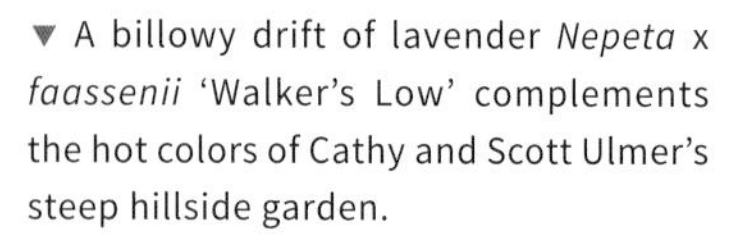

▼ A billowy drift of lavender *Nepeta* x *faassenii* 'Walker's Low' complements the hot colors of Cathy and Scott Ulmer's steep hillside garden.

▶ *Nepeta* x *faassenii* 'Six Hills Giant'.

Polemonium boreale 'San Juan Skies'
Northern Jacob's Ladder

- Full sun
- Fast-draining, moderately fertile soil
- Blooms spring until fall
- Height 10", width 16"
- Zones 4–8

When you need just a little something for the front of your border or want to add a touch of soft color to a rockery, 'San Juan Skies' will do either with aplomb. These enchanting little plants have medium-blue flowers with nicely contrasting white and yellow eyes. They begin blooming fairly early in spring and continue through most of the summer. The flowers sit just above, and nearly cover, a low mound of green foliage. Like many Jacob's ladders, the leaves are small and oval in shape and are arranged in pairs along short stems. Unlike many others of the genus, I've not seen them self-sow.

This Jacob's ladder was apparently discovered growing on a wind-swept rocky outcropping, where it was regularly exposed to salt spray. Though I've never tried them in such conditions, their delicate look clearly disguises a tough and resilient nature.

▲ *Polemonium boreale* 'San Juan Skies'.

▲ Cloaked in soft lavender-blue flowers with intricate centers, the blossoms of 'San Juan Skies' are as attractive to bees as they are to gardeners.

Primula Resources

American Primrose Society
www.americanprimrosesociety

Barnhaven, a nursery specializing in *Primula auricula*
www.barnhaven.com/alpine-auricula

Primrose World
www.primroseworld.com

Primula (revised edition), by John Richards, Timber Press.

The Plant Lover's Guide to Primulas, by Jodie Mitchell and Lynne Lawson, Timber Press

▼ The delicate flowers of *Primula matthioli* are also known as alpine bells. The foliage is light green and slightly crinkled.

Primula
Primrose, Cowslip

A book about cool plants for cold climates would not be complete without a discussion of the incredible genus *Primula*, known affectionately as primroses. Primroses absolutely thrive in relatively cool and moist summer climates with cold winters. With more than 400 options available, they have been organized into groups called sections with the goal of making them easier to study and understand. To some this might just add another layer of complication. Nonetheless, there will surely be a primrose that steals your heart and suits your garden. Several of the real standouts for cold-climate gardeners are described here, but there are so many other options that I am compelled to include photos of a few others. If you find this group of plants to your liking, you can find additional information in books and from societies dedicated to *Primula*.

Most *Primula* prefer humus-rich, moist, well-drained soil. At northern latitudes they can be grown in full sun or part shade. In hot summer areas, ample water and a little afternoon shade is suggested. Their flowers have a delicate freshness that belies their hardy and robust nature. Many varieties also add a seductive fragrance to the garden. Their quiet charm makes them a natural in a woodland garden, along a path, or at the front of a border,

◀ One of the drumstick primroses, *Primula denticulata* 'Ronsdorf Strain' is among the earliest to bloom. As the flowers open, the stem grows longer, holding the "drumstick" well above the foliage, which also grows larger as the season progresses.

▲ Beautiful at the edge of a woodland or a pond, these lovely primroses also bestow an alluring perfume to their surroundings.

where they can be admired closely. Others are at home in a moist setting, while some groups do well in sunny rock gardens.

One attribute of primroses, which you might consider either a benefit or a drawback, depending on your goals, is that they are highly promiscuous. They will freely cross with others of the same or related species. Although this often produces delightful surprises, it can also result in unattractive offspring. If you have a specimen with a particularly cherished color, keep it well away from others of different hues to preserve the blossom color you prefer in the seedlings.

Primula auricula
Auricula primrose, Mountain cowslip, Bear's ear

- Full sun to part shade
- Moist, well-drained, humus-rich, slightly sweet soil
- Blooms in spring to early summer
- Height 5–7", width 5–8"
- Zones 3–8

This primrose has meaty, waxy, compact foliage arranged in a low rosette. It is an early bloomer that does well in sunny rock gardens and other well-drained environments. Flowers are held above the leaves on four- to five-inch stems.

▲ An unusual flower that always causes a stir during garden tours, poker primrose, *Primula vialii*, opens lavender from the bottom, creating a two-toned display. Sadly, it is not as dependable in the colder zones as many other primroses.

▲ Tiny, but always noticed because of its distinctly orange flowers, *Primula cockburniana* is short-lived, so let it self-sow in your garden.

▲ Ronsdorf hybrids are available in a range of colors, including pure white and deep lavender-blue.

▲ Sunny yellow *Primula auricula*.

▲ An exceptionally pretty soft lavender and white auricula primrose. The late Verna Pratt's garden.

▲ An early bloomer, *Primula auricula* mixes well with other spring flowers.

Some varieties are one solid color, whereas others are bicolored in a broad range including red, burgundy, scarlet, lavender, and purple. The centers of the two-toned ones are either light-colored or yellow. Most common are those that are solidly bright yellow or deep purple on the outer edges with a yellow center. Two-toned *Primula auriculas* are quite stunning when the colors are pure and intense. This species also readily crosses in the garden, sometimes producing inferior offspring but occasionally creating a spectacular result. Consequently, it's best to purchase this species in bloom so that you can see whether you like the flower color before you buy it.

▼ *Primula florindae* will spread through seedlings, creating a nice large clump. If you want to manage this expansion, deadhead the flowers when spent.

Primula florindae
Tibetan primrose

- Full sun to part shade
- Moist, well-drained, humus-rich soil
- Blooms mid-summer until frost
- Height 30–36″, width 18″
- Zones 3–8

The largest of all the primroses, *Primula florindae*, hails, as do many tall primroses, from moist meadows in the Himalayas. It is important not to let this plant dry out. In fact, they do best in moist soil and can tolerate occasional flooding. Its bell-shaped and wonderfully fragrant blossoms appear in swirls atop three- to four-foot stems. When a large planting of these near my pond comes into bloom, its perfume fills the air, compelling garden visitors to seek the source of the heady scent. Tibetan primrose is usually available in sulfur yellow but can also be found in a delicate and hauntingly attractive rust color, as well as burnt orange and occasionally even a reddish hue. The rust and orange colors are very special and should be planted separately from the yellow ones to keep the colors pure within their clump. They bloom for an extended time from mid-summer into September.

The large, medium-green foliage is toothed along the edges and is heart-shaped at the base, but still oval in appearance. Another distinguishing feature of this species of *Primula* is its red roots, though few folks ever see them. This plant will self-sow in your garden.

▲ A pretty pink *Primula marginata*.

Primula marginata
Silver-edged Primrose

- Full sun to part shade
- Moist, well-drained, humus-rich soil
- Blooms very early spring
- Height 2–4", width 6–12"
- Zones 3–8

One of the earliest primroses in cultivation, dating from the late 1700s, *Primula marginata* is also one of the earliest to bloom. Its exquisite soft-blue, lavender, or warm-pink flowers are a welcome sight as the snow recedes. The color varies from plant to plant; my favorite is costumed in the purple of wild woodland violas (*Viola labradorica*), a truly "royal" purple. On the other hand, it's hard to resist the hauntingly lovely periwinkle blue ones, either. These tiny plants are best enjoyed up close, where you can see the delicate pale, pinking-sheared edges of their small, leathery leaves, usually covered on both sides with a light frosting of white farinose (the mealy powder coating that naturally occurs on many different primroses). Being part of the auricula section, they do well in rock garden settings as well as in regular garden soil, so long as the drainage is excellent. Clumps expand slowly and are easily divided. They are lightly fragrant.

▲ Silvery edges that look as if they've been trimmed with pinking shears give this plant one of its common names: silver-edged primrose. I can't resist the hauntingly beautiful violet-blue of its flowers.

Primula waltonii and *Primula alpicola*
Asiatic primrose

- Full sun to part shade
- Moist, well-drained, humus-rich soil
- Blooms mid-summer
- Height 18–24", width 12–15"
- Zones 3–8

Two favorites among the Sikkimensis section are *Primula waltonii* and *Primula alpicola*, both of which are also sweetly fragrant and have bold rosettes of medium-green, oblong-shaped leaves. They bloom a bit earlier than *Primula florindae*, usually beginning in late June. There are very subtle differences in

▲ *Primula waltonii*.

▲ Just-opened blossoms face upward but then turn away shyly as more flowers open on the stem.

▲ *Primula alpicola*.

▲ A wonderful mixture of tall primroses in my garden. Their fragrance is divine.

the details of the leaves of these two species and the amount of farinose on the flowers and stems; however, these differences are generally inconsequential. Flower stems vary in height between twelve and twenty inches. *Primula waltonii* flowers will most likely be strawberry to wine red, whereas *Primula alpicola* blossoms will usually be white, cream, lavender, or purple. Once mixed in the same area, though, even these distinctions will soon disappear among the off-spring, because these primroses will readily cross. In one area of my garden, I've let a group of these intermingle. The variation in the resulting plants has made a delightful medley of color.

Primula x *juliae*
(synonym *Primula* x *pruhoniciana*)
Juliana Hybrids

- Full sun to part shade
- Moist, well-drained, humus-rich soil
- Spring-blooming
- Height 6", width 12"
- Zones 4–9

One of the ancestors of these hybrids, *Primula juliae*, was discovered in the eastern Caucasus in the early nineteenth century. Intense interest in this delightful new find led to enthusiastic hybridizers crossing it with many other primroses. The most famous result of this work is *Primula* x *juliae* 'Wanda', winner of the Award of Merit in 1919. Its vibrant wine-colored blossoms with a touch of yellow

▲ *Primula* x *juliae* 'Wanda'. Marguerite and Flip Felton's garden.

▲ *Primula* x *juliae* 'Dorothy'.

at the center are tucked in among well-toothed, apple-green, oval leaves. It blooms prolifically from early spring into early summer.

Another spring-blooming and incredibly floriferous Juliana hybrid is *Primula* x *juliae* 'Dorothy'. 'Dorothy' holds her pale yellow flowers just above the foliage rosette, nearly obscuring it with an incredible mass of blooms. She has a noticeable dark yellow eye at the center of each flower.

There are many other Juliana hybrids to choose from, but these two are exceptional and, like the others of this clan, make excellent garden plants. Their clumps will expand nicely with time and are easy to divide after blooming is finished.

Pulsatilla vulgaris 'Violet Bells' and 'Red Bells'
Pasque Flower

- Full sun to part shade
- Fast-draining, moderately fertile soil
- Blooms in very early spring, followed by attractive persistent seed heads
- Height 10–12", width 12–18"
- Zones 3–7

In northern gardens *Pulsatilla vulgaris* emerges as the snow begins to melt away. These colorful plants are not fazed by late frosts or even a late snow, so if it's early color you crave, consider adding these to your garden. Though the species is medium purple and quite lovely, I've found 'Violet Bells' to be somewhat darker and larger. 'Red Bells' gives those who favor red a great choice in an attractive warm hue, similar to that of merlot wine. Such large flowers on a diminutive plant create an outsized impact.

▲ *Pulsatilla vulgaris* 'Violet Bells'.

▲ *Pulsatilla vulgaris* 'Red Bells'. Gari and Len Sisk's garden.

Each cup-shaped blossom has six hairy, slightly pointed sepals (rather than true petals) surrounding a mass of prominent golden stamens. The vibrant combination of deep purple with gold, or wine-red with gold, looks marvelous with early spring bulbs or yellow *Primula auricula*.

Foliage is finely divided, feathery in appearance, and somewhat hairy. It forms a low, light green mat before solitary flowers appear on individual stems holding each bloom a few inches above the leaves.

After flowering finishes, wonderful fuzzy seed heads, similar to those of many clematis, appear atop the flower stems. At this point, these stems grow a bit, reaching more than twelve inches. The fluffy pompoms persist for several months and provide one more spritely element to use in combinations. All in all, this is a cool plant. It will do fine in a well-drained mixed border but will excel in a rock garden.

Rheum palmatum 'Atrosanguineum'
Ornamental Rhubarb

- Full sun to part shade
- Moist, humus-rich soil
- Interesting all season
- Foliage height 42", flower height 8', width 6'
- Zones 3–8

Ornamental rhubarb dons so many costumes during the course of a season that it's difficult to select a representative photograph. As the garden's first bulbs emerge in early spring, *Rheum palmatum* unfurls its massive brilliant rose leaves to herald the new season. This rosy beginning foretells a bit about its

▲ In spring, as the colorful leaves of 'Atrosanguineum' unfurl, they are dark on both sides.

▶ Later, the top sides will become primarily green while retaining their intense color on the underside. Note the white flowers in this example.

▶▶ Even in a lushly planted mixed garden like the one I designed for the Homer Garden Club's garden at the Baycrest Overlook, *Rheum palmatum* 'Atrosanguineum' is a stunning focal point.

finishing flourish, for the eight-foot-tall flower stalks that will appear in late summer are usually robed in the same hue. Between these two events, the foliage transitions to a rich, burgundy-infused green above with dark burgundy on the reverse. Even a slight breeze will flutter the two- to three-foot deeply cut leaves, creating a dazzling effect. This plant is a natural as a focal point; surrounded by bold neighbors, it will still hold center stage. Few plants that enjoy full sun have such large foliage, making this one a designer's dream.

At maturity, *Rheum palmatum* 'Atrosanguineum', the best of the cultivars, can exceed six feet in width, so give it plenty of room. I've described the flower stalks as *usually* being robed in the same hue as the early foliage, but there is a good deal of variation among plants sold under the same name. Unfortunately, in some, the foliage becomes almost entirely green as the season progresses; in others, the foliage is divine, but the flowers are white or pink. This unpredictability has resulted in fewer nurseries offering this plant. Nonetheless, if you can find one with the rich coloration, it will be a star in your garden.

Rodgersia
Roger's Flower

This is a majestic genus with incredibly beautiful foliage—big, bold, colorful, well textured, and distinctively shaped. Its members are all moisture-loving and do particularly well along a stream bank or the margins of a pond. They are good candidates for rain gardens provided they don't dry out for too long. Mainly found in woodland settings in Asia, *Rodgersia* are finally getting the attention they deserve in North America. Though you'll see them described as Zone 4, 5, or even 6, my view differs. If their horticultural needs are well met, they are clearly hardy to Zone 3. The proof of this can be found at the Alaska Botanical Garden in Anchorage, Alaska, a very chilly Zone 3 location with protracted winters. There, many healthy clumps of different species have done exceedingly well for years. It's best to keep these large-leaved beauties in locations that are sheltered from wind.

Rodgersia podophylla 'Rotlaub'
Bronzeleaf Rodgersia, Redleaf Rodgersia

- Sun in far northern latitudes, part sun otherwise
- Richly fertile, moist soil
- Blooms in summer
- Height 32–36", width 3–4'
- Zones 3–9

▲ *Rodgersia podophylla* 'Rotlaub' in its cool weather color. Gari and Len Sisk's garden.

▲ As the season progresses, 'Rotlaub' foliage exhibits more green, but in the fall, the intensity of bronze returns.

A superb specimen plant to use as a focal point or in combinations in a mixed border, this *Rodgersia* emerges saturated bronze-red and matures to a stunning dark bronze. The color intensifies in the fall. Each shiny leaflet is distinctly veined and has jagged, finely toothed outer edges; they narrow toward the center where they attach to their tall stem. The overall effect is of a pretty parasol.

In mid- to late summer this fantastic foliage is augmented by a dramatic plume of white flowers. Small, star-shaped blossoms are arranged on tall panicles, somewhat reminiscent of astilbe blooms, that rise dramatically above the foliage. It's best to cut back the flower stalk after blooming is complete to direct energy back to the plant.

Rodgersia henrici 'Cherry Blush'
Fingerleaf Rodgersia

- Sun in far northern latitudes, part sun otherwise
- Richly fertile, moist soil
- Blooms in summer
- Height 32–36", width 3–4'
- Zones 3–9

This is another candidate that is well suited to play a starring role in your garden. Deeply textured, quilted-looking, ovate foliage unfurls coppery-red in early spring and holds its deep coloration through much of the season, especially in cool summers. As leaves reach maturity, they turn greener in the center with red along the outer edges. As weather cools in fall, 'Cherry Blush' foliage is again infused with more red coloration.

▶ This newly planted *Rodgersia henrici* 'Cherry Blush' will soon hold its beautifully textured foliage high above the surrounding plantings.

▶▶ A *Rodgersia* sp. grown from seed gathered in Yunnan, China, puts forth blossoms that stand tall on colorful burgundy stems. Though the species of this specimen is unknown, you can achieve a similar look with *Rodgersia aesculifolia.* Teena and Peter Garay's garden. Teena's design.

Scarlet red stems rise elegantly above the wonderful foliage to support a mass of small deep rose flowers arranged along the stem. The color of the stems and flowers reminds me of maraschino cherries.

Rodgersia astilboides
(As with many plants, this one has been recently moved to a different genus. So much for botanical nomenclature clarifying things! You will find it described under *Astilboides tabularis.*)

Salvia nemorosa
Garden Sage

There are more than 700 species within the genus *Salvia.* Some are known for their culinary uses, while others are more ornamental. Many commonly known sages are brilliant red annual bedding plants; still others are tender perennials. *Salvia nemorosa* is one of the hardiest species. Like all members of the mint family, it has square (four-sided) stems with whorls of flowers adorning its upright spikes. When crushed the foliage releases a spicy fragrance. It tolerates a wide range of soil pH, from acidic to neutral to sweet.

▼ *Salvia nemorosa* 'May Night' presents long-blooming spikes of dark blue flowers.

Salvia nemorosa 'May Night'
Wood Sage

- Full sun
- Moderately fertile, moist, but well-drained soil, acidic to alkaline
- Blooms spring and summer
- Height 18–20", width 18–24"
- Zones 3–8

'May Night' is one of the most dependable salvias for cold climates. A profusion of upright spikes of dark violet-blue dazzle nonstop for months on end, even longer if you deadhead. Use this saturated color as an excellent foil for white, yellow, orange, even light blue or lavender. In fact, because 'May Night' blooms for so long, you could design a succession of color combinations around it just for the fun of it.

The plant forms a dense clump that is drought-tolerant once established. Its leaves are oblong, gray-green, and strongly fragrant. In your garden, it will delight not only you but also bees, butterflies, and hummingbirds.

Resources for Saxifraga

The Saxifrage Society, an international organization
www.saxifraga.org

Saxifraga x *arendsii*

Moss Saxifrage, Mossies

- Full sun to part shade
- Moderately fertile, moist, well-drained, lightly acidic to alkaline soil
- Spring bloomer
- Height 6–9", width 12"
- Zones 2–7

Though saxifrages have a reputation—well deserved in some cases—for requiring precise growing conditions, the hybrid *Saxifraga* x *arendsii* is an exception. It will grow in slightly acidic, neutral, or alkaline soil. We often think of saxifrages as rock garden plants—and they do perform exceptionally well in the fast-draining environments of both rock and crevice gardens—but this one will also do just fine in well-drained garden soil with plenty of humus.

Some of the species in the heritage of *Saxifraga* x *arendsii* hail from north of the Arctic Circle, so this sweet little specimen is hardy enough for most

◄ *Saxifraga apiculata* 'Gregor Mendel'. The Alaska Botanical Garden.

▼ White blossoms with lime green centers dance above the mat of saxifrage foliage.

▼ Pretty in pink, this "mossy" is tucked between rocks in a large rockery. Notice how tight the foliage looks as a result of its snug niche. Mary and Karl Schneider's former garden.

gardeners of the extreme north. What will quickly lead to its demise, though, is a hot, humid summer climate or drought. About forty cultivars are available, and frankly, I cannot tell one from the other without a label except to say that one is white, another pink, and a few rose, carmine, or deep red. 'Triumph', one of the dark reds, is a particularly successful cultivar. All the "mossies," as this group of saxifrages is called, have tiny star-shaped and cupped flowers. In spring, their perky presence, held aloft on thin stems above a mat of mossy looking foliage, demands attention. Tucked between rocks to keep the mat tight or cascading over a rock wall, they are a very welcome sight in spring.

There is a myriad of other saxifrages, many of them best suited to a very fast-draining rock garden. A delightful, pale yellow one with outsized flowers, *Saxifraga apiculata* 'Gregor Mendel', can be seen in the rock garden at the Alaska Botanical Garden in Anchorage.

Sedum
Stonecrop, Sedum

The wide range of sizes, shapes, and growth habits of *Sedum* offers us such variety that there is a sedum for nearly any sunny garden niche. These drought-tolerant, fleshy succulents are fabulous in rock gardens, provide interesting textural contrast in a mixed border, and do well in specialty applications such as green roofs and walls. Most references indicate that stonecrops do best in neutral to alkaline soil, but my experience is that they do perfectly well in mildly acidic soils, too. They can be annual, biennial, or perennial. The ones described here are all perennial and hardy.

Because of its interesting and unique texture, the foliage of this genus is appealing from spring snowmelt until after snow returns in the fall. *Sedum* flowers typically have five petals and are star-shaped. Most bloom from mid-summer into fall, so they provide wonderful late-season beauty for us and sustenance for pollinators that are frantically preparing for winter.

Although I am always surprised when I hear someone say that he or she rather dislikes stonecrops, I've chalked it up to the excessive use and rambling, somewhat floppy nature of *Sedum kamtschaticum*. Its appearance in a small pot at the nursery belies its true nature in the garden, where it often disappoints. As luck would have it, there are many species and cultivars that are much more interesting. In fact, I've had a difficult time limiting the number presented here. From that statement, you may accurately deduce that I find these tough little plants not only cool, but also fascinating.

▲ *Sedum* is available in a wide variety of sizes and shapes. There are some fantastic flower forms as well.

▲ Sedum is unphased by a light snow and will continue blooming well into autumn.

In his excellent book, *Herbaceous Perennial Plants: A Treatise on Their Identification, Culture, and Garden Attributes*, Dr. Allan M. Armitage writes on pages 918 and 919 that "finding the correct name for a stonecrop is an adventure in frustration" and then gives this good advice: "Do not lament if you're unsure of the botanical name of a stonecrop in the garden, you are in good company." I quote Dr. Armitage to give you comfort—and also to beg for forgiveness if you think I've misidentified one of these selections. Adding to the overall confusion surrounding this genus, a whole raft of them has been moved to the genus *Hylotelephium*! Because virtually no nurseries have adopted the new names, and to keep things simpler, I'll continue to use their historical designation: *Sedum*.

Sedum album 'Coral Carpet'
Coral Carpet Stonecrop

- Full sun or part shade
- Moderately fertile, well-drained soil
- Blooms summer to fall, foliage interesting all season
- Height 1–3", width 12–24"
- Zones 3–9

Descriptively named, this small-leaved, ground-hugging cultivar has green foliage that turns to a pretty coral in cool weather. Topped by a mass of palest pink flowers, the combination is immensely attractive. It does exceedingly well in rock gardens, where it will snake between large rocks, creating the impression that it's flowing downhill. Little pieces that break off will root and expand the clump if you let them. Otherwise, pluck them out and share them with a friend. 'Coral Carpet' can be vigorous in the perfect environment, so site it away from tiny, slow-growing alpines.

▲ *Sedum album* 'Coral Carpet'.

Sedum spurium 'Blaze of Fulda'
Blaze of Fulda Stonecrop

- Full sun or part shade
- Moderately fertile, well-drained soil
- Blooms summer to fall, foliage interesting all season
- Height 3–6", width 10–12"
- Zones 3–9

If you enjoy bright colors, this is the low-growing sedum for you. From midsummer until the snow flies, small clusters of cerise flowers appear on short deep pink stems with rose leaves along their length. The foliage mat below the flowers is mostly green to bronze, with reddish edges. These colorful, fleshy leaves are arranged in swirls that look like flowers themselves. Not only is this selection

◄◄ Brilliantly colored 'Blaze of Fulda' is a rock garden star.

◄ *Sedum cauticola* in its spring finery.

▲ By fall, bright pink flowers join the show and the succulent leaves take on a warm rose glow.

fabulous in a rockery, but it also makes an excellent container plant, bringing color and texture to your combinations all season. As with all sedums, bees and butterflies find it irresistible.

Sedum cauticola
Cliff Stonecrop

- Full sun or part shade
- Moderately fertile, well-drained soil
- Blooms summer to fall, foliage interesting all season
- Height 2–3", width 10"
- Zones 3–9

Gray-blue foliage tinged with pink, stems flushed with purple, and pink-purple flowers that age to carmine combine to create a treasure trove of delicate hues. This is a lovely choice to blend with pastels or to contrast with burgundy or saturated purple. Like all stonecrops, the foliage of this plant offers nice texture and color all season. The flowers join the show later but carry on until the garden is covered by a blanket of snow. Each gray-blue leaf is round and edged in pink. Like other ground-hugging sedums, the leaves swirl around the stems, creating the look of floral rosettes. It's captivating.

Sedum spectabile 'Autumn Fire'
Autumn Fire Stonecrop

- Full sun or part shade
- Moderately fertile, well-drained soil
- Blooms summer to fall, foliage interesting all season
- Height 24–30", width 18–24"
- Zones 3–9

▶ A fabulous tall *Sedum*, 'Autumn Fire' stands boldly upright, holding incredibly large flower clusters aloft.

▶▶ As the pink buds open, they reveal rose-colored blossoms.

'Autumn Fire' appears similar to the immensely popular 'Autumn Joy' but is a little shorter and has stouter stems. The combination of these differences makes 'Autumn Fire' more reliable at staying erect throughout the season. If you want an even shorter result, however, you can pinch the plants back early in the season. I like to pinch back just the stems toward the viewing side of the clump to create a cascade of blooms. Bright, rose-colored flowers adorn the tops of stems costumed in very pretty mid-green, fleshy foliage. As with other tall sedums, don't cut them back in fall or you'll miss a wonderful part of the show. The birds will be better fed, too!

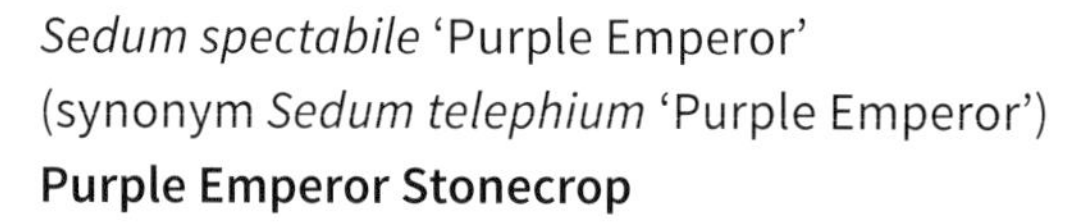

Sedum spectabile 'Purple Emperor'
(synonym *Sedum telephium* 'Purple Emperor')
Purple Emperor Stonecrop

- Full sun or part shade
- Moderately fertile, well-drained soil
- Blooms summer to fall, foliage interesting all season
- Height 14–16", width 15"
- Zones 3–9

▲ *Sedum spectabile* 'Purple Emperor'. Denice and Roger Clyne's garden.

Upright sedums are the princely members of this vast genus. Their elegant stature, long season of beauty, and late-season staying power make them a welcome cast member in a mixed border. 'Purple Emperor' adds incredibly dark purple, almost black foliage throughout the season. The richness of this color can't be exaggerated and is even deeper in full sun. As summer progresses, large dusky-pink flowers form at the top of each stem, which grows as the pink flowers open. With age, flowers transition to bronze. Enjoy these plants well into winter: they will stay erect unless a wet snow weighs them down. Seeds provide food for birds in winter. Sedums also make interesting cut flowers, adding an unexpected texture to bouquets.

Tanacetum coccineum 'Robinson's Hybrid'
(synonym *Pyrethrum coccineum*)
Painted Daisy

- Full sun
- Well-drained, moderately fertile or sandy soil
- Blooms mid-summer, will rebloom if cut back
- Height 26", width 14–16"
- Zones 3–7

Robinson's Hybrids are available as a mix, but because I prefer to control my color combinations, I opt for either 'Robinson's Red' or 'Robinson's Pink', depending on my mood or color scheme. The red cultivar is more crimson than true red; it will clash with orange-based reds. Both cultivars sport large, mostly single daisy-style flowers with bright yellow central disks. Some blossoms have so many petals that they appear semi-double. All are classic, old-fashioned flowers with one huge difference: they don't run rampant through your garden scattering seeds hither and yon. I've grown them for nearly twenty years in my Alaska garden and have yet to find a seedling. However, you can propagate them by division very easily.

Painted daisies are long-lasting as cut flowers. They attract butterflies, which find their flat topped flowers a stable and comfortable place to alight while they feed. You can choose to deadhead each daisy as it fades or can wait until the show is nearly over and cut back the entire plant to the basal (bottom) foliage. Don't cut back too soon, though, since the variety of hues produced by the mix of fresh and maturing blossoms provides nice variation. The foliage of these happy-faced plants is green, ferny, and texturally very attractive. It yields a nice contrast when paired with spiky iris foliage or bold-leaved plants. Though dependably hardy, these plants do not fare well in hot, humid climates.

▲ 'Robinson's Red' shines in front of *Artemisia ludoviciana* 'Valerie Finnis'.

▲ For soft pastel combinations, consider *Tanacetum coccineum* 'Robinson's Pink'.

Thymus
Thyme

There are countless puns associated with the word thyme. I'll spare you those, but I do have to share a fact I find incredibly interesting. Not only has thyme been around for millennia, but it was also used by the ancient Egyptians for embalming. Isn't that cool? Or dreadful? Whichever way it strikes you, feel free to use this tidbit to entertain your friends at the next dinner or garden party you attend. Although thyme has long served culinary, medicinal, and ornamental purposes, because this book is about great garden plants, we'll focus on ornamental varieties.

▲ 'Highland Cream' in pink garb mingling with *Ajuga*. Marguerite and Flip Felton's garden.

Most of the 350 or so varieties of thyme are woody-based, evergreen, and, perhaps most notably, aromatic. Many hug the ground and work well as "steppables"—plants that look great between stepping stones and that can even serve as an alternative to grass in low-traffic areas. Each step releases a whiff of fragrance, which varies by variety. They have small, usually oval leaves and produce two-lipped flowers that are packed with nectar. Bees swarm them. Plants need good drainage and little else. They'll grow well in hot, arid locations but don't do well in wet, humid conditions.

Thymus praecox 'Highland Cream'
Creeping Thyme

- Full sun
- Well-drained, slightly acidic, neutral to alkaline soil
- Blooms in summer
- Height 1–2", width 8–18"
- Zones 3–8

This English import has wonderful foliage. Each tiny leaf has a green center with variable splashes of creamy yellow along the edges. From a distance, it appears to be soft yellow. Up close, the variegated pattern is quite interesting and attractive. In late spring and throughout summer, the mat of foliage is covered with flowers that can vary from white to lavender depending on the individual plant. Although this variance may be disconcerting, especially if you want to be in total control of your color scheme, the hues blend well. I have found this serendipitous trait to serve as a refreshing surprise now that I know it might occur, but the first time it happened to me, it was a surprise indeed!

▲ *Thymus praecox* 'Highland Cream' with white flowers and an interesting black-and-white visitor. Marguerite and Flip Felton's garden.

Thymus pseudolanuginosus
Woolly Thyme

- Full sun
- Well-drained, slightly acidic, neutral to alkaline soil
- Blooms in late summer
- Height 1–2", width 12–36"
- Zones 3–8

It's hard to imagine a large rockery without at least some woolly thyme. Incredibly soft, fuzzy gray-green foliage feels as pleasing as it looks. Both the visual and tactile texture of this plant is a welcome contrast to the unyielding solidity of stone. It's the best cascading form of the creeping thymes and will drape several feet over the edge of a wall or spill over rocks or down a stairway. If it gets out of bounds, simply hold the end and cut off the amount you choose. Lavender

▲ *Thymus pseudolanuginosus* cascades over the rocks in Marguerite and Flip Felton's wonderful rock garden.

flowers appear sporadically but don't completely cover the foliage. In this case, that's probably a good thing, because the foliage is so delightful.

▲ The incredible hue of 'Coccineus' steals the show every time. Cheryl and Cliff Schaeffer's rock garden.

Thymus serpyllum 'Coccineus'
(synonym *Thymus coccineus*)
Red Creeping Thyme

- Full sun
- Well-drained, slightly acidic, neutral to alkaline soil
- Blooms in summer
- Height 2–4", width 12–18"
- Zones 3–8

This beauty is a bit too "tall" to be used as a steppable. It forms a low, dense mat of dark green foliage. From mid-spring through most of summer, the foliage is hidden beneath a deep blanket of brilliant magenta flowers. This incredible color adds enormous pizzazz to a garden. Try it cascading between dark colored rocks or spilling over a wall. Unless you prefer quiet colors, you're going to love this easy-care thyme. Butterflies and bees love it, too. After flowers are finished and the weather cools, the mat of foliage turns a striking bronze.

▲ Rose-colored buds open pale, pale pink on the ground-hugging mat of 'Pink Chintz' foliage.

Thymus serpyllum 'Pink Chintz'
Creeping Thyme

- Full sun
- Well-drained, slightly acidic, neutral to alkaline soil
- Blooms in summer
- Height 2–4", width 12–18"
- Zones 3–8

When this creeping thyme is partly in bud and partly in bloom, it's at its prettiest. Buds are deep rose, slowly opening to pale pink flowers. I find this long-lasting combination very appealing. After all the buds open, the slightly hairy, green mat of foliage is completely hidden by an expanse of soft pink blossoms. When not in bloom, the soft, fuzzy leaves are a useful textural element that can serve as a background to focus attention on a more dashing cast member in your garden production.

Trollius
Globeflower, Trollius

Globeflower foliage in nearly all cases is dark green, deeply and palmately lobed (five-fingered like a hand), with toothed edges. Though not well-known or successfully grown south of Zone 7, bright yellow *Trollius europaeus*, common European globeflower, is well loved and somewhat ubiquitous in

▲ Trouble-free *Trollius europaeus* is often seen in spring in cold-climate gardens.

▲ *Trollius chinensis* 'Golden Queen' is taller and later than most of its relatives. It is joined in this image by a backdrop of *Astrantia major* 'Hadspen Blood'. Homer Garden Club's Baycrest garden.

▲ Early-blooming, diminutive *Trollius pumilus* adds a bright punch to the spring garden, but both its flower form and foliage are quite different from those of most other globeflowers.

Alaska and other moist, cold climates. What are less well known, even in the north, are some of the other species and interesting cultivars. Among these are a few that are ivory or orange. Let's take a look at them.

Trollius chinensis 'Golden Queen'
Queen of the Buttercups, Globeflower

- Full sun to part shade
- Fertile, humus-rich, moist acidic, neutral, or alkaline soil
- Blooms mid- to late summer
- Height 30–36", width 18–24"
- Zones 3–7

The flower form of this cultivar is different from that of most globeflowers. Bright tangerine sepals form an open cup revealing upright, narrow petals that look for all the world like a crown. These fascinating flowers are held on slim, leafless stems eighteen inches or more above a mound of rich green foliage. This cultivar blooms later than most globeflowers, withholding its blossoms from the garden until mid to late summer. Use this to your benefit if you are a fan of *Trollius* since its delayed bloom will help you ensure that there is a trollius of some sort adding color to the garden throughout the season. Foliage is attractive all summer and is bold enough to contrast with fine or spiky companions.

Trollius pumilus
Dwarf Trollius, Dwarf Globeflower

- Full sun to part shade
- Fertile, humus-rich, moist acidic, neutral, or alkaline soil
- Blooms mid-spring to early summer, second bloom late in summer
- Height 8–12", width 10–12"
- Zones 3–7

A delightful petite plant for the front of the border or tucked into a rock garden, *Trollius pumilus* packs a punch with its bright yellow flowers and glossy dark green foliage. Five-petaled flowers open flatter than, but in the same classic yellow as, common globeflower (*Trollius europaeus*). They're held just above a compact, round clump of foliage. Each leaf is made up of five lobes that are further divided into three parts, creating an attractive frilly look that adds nice texture to the garden.

Dwarf trollius blooms for about six weeks in cool weather. Cut back the spent flowers to encourage a second round of blossoms. This plant will self-sow close to the parent plant; seedlings can be left to increase the size of the planting or easily removed to share with others.

Trollius x *cultorum* 'New Moon'
New Moon Globeflower

- Full sun to part shade
- Fertile, humus-rich, moist acidic, neutral, or alkaline soil
- Blooms all summer
- Height 18–24", width 12–18"
- Zones 3–7

▲ The buds of beautiful ivory-colored 'New Moon' are somewhat darker than the flowers. Homer Garden Club's Baycrest garden.

There are several cream-colored cultivars available, including the wonderfully named *Trollius* x *cultorum* 'Alabaster'. Though I enjoy that cultivar in my own garden, I've found 'New Moon' to be much easier to establish, being a bit larger and more vigorous than 'Alabaster' or any of the other pale selections. Its lovely double flowers have an incredible soft, creamy tone; it's otherworldly. Buds sheathed in soft rose, as well as dark green, spiky seed heads, join the pretty flowers to make the entire composition especially attractive. Blooms are held high on long stems well suited to cutting for bouquets.

As with other varieties of globeflower, the foliage of this cultivar is five-lobed, deeply cut, and a rich dark green. Because it is both mid-sized and well textured, it makes an excellent companion for plants with bold or strappy foliage as well as for those selections with fine or frilly leaves.

Veronica spicata 'Royal Candles'
Spiked Speedwell

- Full sun
- Moist, well-drained, loamy, moderately fertile, acidic to alkaline soil
- Blooms early summer to early fall
- Height 10–16", width 10–15"
- Zones 3–8

▲ 'Royal Candles' produces masses of spiky flowers that add color and form to the garden for most of the summer. Homer Garden Club's Baycrest garden.

Midnight blue spikes of flowers adorn this incredible performer from early summer until early fall. Not only is that an amazingly long period of bloom, but its impact is made even greater because the flower spikes (called racemes) make up half the height of the plant. Silvery, pale-green buds open into individual dark blue flowers from the bottom of each pyramid-shaped raceme, creating a pleasing two-toned appearance.

Adding to the overall spiky architecture of this plant are shiny, dark green, lance-shaped leaves that are toothed along the sides but not at the pointed tip. The compact nature of 'Royal Candles', along with foliage that is attractive all the way to the ground, usually makes this an excellent front of the border or edging plant, though on occasion it suffers from powdery mildew late in the season. This gem will also look great in a container, where its season-long bloom rivals annuals. All in all, this is a fabulous selection, the nectar of which attracts butterflies by the dozen.

Veronicastrum virginicum 'Apollo'
Culver's Root

- Full sun
- Moderately fertile, humus-rich, moist soil
- Blooms mid-summer to fall
- Height 60–66", width 42–54"
- Zones 3–8

▲ Late-blooming *Veronicastrum virginicum* is a magnet for bees late in the season.

With such a tall, commanding presence, this is an excellent choice for the back of a border. Nine-inch-long elegant spires of lavender-blue flowers rise atop unbranched, upright stems cloaked in whorls of rich green foliage. Leaves are lance-shaped and toothed along the edges. It is the arrangement of the leaves that primarily differentiates *Veronicastrum* from *Veronica*. The latter has either alternate or opposite leaves rather than having them all around the stem. Nonetheless, the folks who make decisions about which plants belong to which genus have joined them together from time to time. Nurseries tend to stick with the designation used here, but you may also find them listed under *Veronica*.

If your season is long enough, just as the central raceme of flowers reaches its peak, side shoots will emerge to continue the show. These are shorter and surround the main flower spike, creating a candelabra of flowers. Some folks cut the center flower stem to force this to happen earlier. Either way, pollinators, especially bees, are drawn in droves to this late bloomer.

Though many references say that these flowers will do well in part shade as well as full sun, it's been my experience that they will not. With insufficient sun, they tend to reach for the light, often culminating in an early demise. My advice is to give them lots of sun and space. Because the foliage on the lower parts of the stems discolors and withers in an unattractive way before the plant has finished blooming, it is best to site any culver's root, including 'Apollo', in the back of the border with a group of shorter, leafy plants in the foreground.

Zigadenus elegans (synonym *Anticlea elegans*)
Death Camas, Camas Wand Lily

- Full sun to part shade
- Fertile, well-drained soil
- Summer
- Height 24–30", width 10–16"
- Zones 4–9

For those of you who find flowers tinged with green to your liking, *Zigadenus elegans* should hold great appeal; it is certainly one of my favorites. Tall, upright stems with small leaves partway up support a mass of flowers three-eighths of an inch wide on approximately inch-long side shoots. The short side stems are covered in a blush-colored, papery sheath and grow closer together and more numerous as they approach the top. Attractive, slightly cupped, upward-facing blossoms have six white petals each with a green heart-shaped gland at the base. Seen from a distance, the overall hue appears pale chartreuse. The interesting flowers have a waxy appearance and persist for more than a month and a half. Rich green gently arching, basal foliage has a crease down the middle of each leaf as if it had been folded at some point. It surrounds the cluster of flower stems, creating a charming "bouquet."

▲ Unusual but spectacular *Zigadenus elegans* blooms prolifically when grown in a well-tended garden. An Alaska native, it is much less floriferous in the wild.

Death camas is an easy-care plant. It can be deadheaded after blooming if you wish, but its seed heads stay nice for a prolonged period, turning chocolate brown with time. It doesn't appear to ever need dividing; the clump in the photo has been in my garden without division for fourteen years. On the other hand, should you choose to divide your clump to create additional plants, this is easy to do. Dig up the entire plant, and gently pull the bulbs apart.

The common name, death camas, derives from the toxicity of *Zigadenus* combined with the similarity of its foliage to that of *Camassia quamash* (wild hyacinth), an edible plant and food source for Native Americans. Apparently, the bulbs of wild hyacinth were harvested before the differentiating flowers bloomed and were then roasted in wet leaves over an open fire. Digging and eating the wrong plant could be fatal. Because *Zigadenus* is so toxic, owing to a high concentration of alkaloids, and because the foliage is difficult to tell apart from that of wild hyacinth, it is said that the women of the tribes were sent into the fields during bloom periods to remove any plant with the pretty green-tinged flowers. The men were left to hope that the women hadn't missed any!

▲ Each blossom has a small, heart-shaped green gland at the base of every petal.

Although native to a wide swath of North America, including Alaska, this plant is rarely found in nurseries. Your best bet is to beg a divide from a friend who has it or to visit the native plant and seed suppliers found online.

SHRUBS

Shrubs are woody perennials. Their framework or structure remains in place throughout the year, adding dimension, mass, texture, and color to a garden at all times. Deciduous shrubs lose their leaves during the cold seasons; evergreens do not. The best ones offer spring or summer flowers, attractive foliage throughout the growing season, and brilliant colors in the fall. Often their bark will add a touch of color to the winter garden as well.

Berberis thunbergii
Japanese Barberry

If you value colorful foliage as I do, you'll find a lot to appreciate in *Berberis thunbergii*, the most popular of the cold-climate barberries used for landscaping. The species is dense and round in silhouette, with leaves that are green on top and bluish green on the underside. The foliage turns rich shades of red and orange in the fall, making it particularly lovely at that time of year, but it is the cultivars and varieties that have set the gardening world abuzz. Because of their dense nature and abundant thorns, the larger barberries make great shrubs for barrier hedges. The smaller ones and many cultivars are welcome guests in mixed borders. They leaf out early, an important attribute to gardeners with short seasons. These are low-maintenance shrubs that can be pruned at any time. One word of warning: be careful when gardening near them; the thorns are sharp!

Berberis thunbergii 'Aurea'
Golden Barberry

- Full sun to part shade (fruiting and fall color are better in full sun)
- Tolerant of most well-drained soils, regardless of pH or fertility
- Foliage attractive all season, flowers in spring
- Height 36–48", width 36–48"
- Zones 4–8

◄ Luminous and eye-catching, golden barberry puts on a vibrant performance in this nook in Marilyn and Sam Beachy's garden; Gee Denton's design.

Although we most often envision burgundy or purple foliage when we think about *Berberis thunbergii*, there is also a dazzling golden yellow cultivar called 'Aurea'. It is a bit more upright than some and has a looser form. Combined with one of its purple or burgundy mates, this shrub is simply eye-popping. Its small oval leaves, which turn pretty shades of orange and red in fall, complement and contrast wonderfully with the red or burgundy glow of many of their purple relatives. Because of the intensity of its color, you might wish to use it sparingly, but it will definitely draw the attention of garden visitors. Yellow flowers are barely visible against the leaves, but red berries can add a bright note of interest as autumn approaches.

▲ Richly-colored and compact, 'Crimson Pigmy' is the perfect choice for small gardens. Homer Garden Club's Baycrest garden.

Berberis thunbergii var. *atropurpurea* 'Crimson Pygmy'
Crimson Pygmy Barberry

- Full sun to part shade (fruiting and fall color are better in full sun)
- Tolerant of most well-drained soils, regardless of pH or fertility
- Foliage attractive all season, flowers in spring
- Height 18–24", width 36–42"
- Zones 4–8

One of the most brilliant of the so called purple-leaf barberries, 'Crimson Pygmy', glows crimson-burgundy. Its rich, lustrous hue is particularly appealing in spring and summer and then, in the fall, turns a handsome intense burgundy. Small yellow flowers form along the underside of the stems in the spring but are not the main attraction. It's the densely arranged, small, oval, colorful foliage that earns the kudos. Like most barberries, it has sharp thorns, making it undesirable to roaming, hungry ungulates. Those spines are also the reason why I cannot attest to what is reported to be a nice flower fragrance. I choose not to put my nose close enough to check!

The small size of this selection makes it ideal for smaller gardens, but it is equally lovely near the front of a large bed or used as a mass planting. It has received the Award of Merit from the Royal Horticulture Society.

▲ *Berberis thunbergii* var. *atropurpurea* 'Crimson Velvet' foliage changes from smoky-purple to deep red in the fall.

Berberis thunbergii var. *atropurpurea* 'Crimson Velvet'
Crimson Velvet Barberry

- Full sun to part shade (fruiting and fall color are better in full sun)
- Tolerant of most well-drained soils, regardless of pH or fertility
- Foliage attractive all season, flowers in spring
- Height 54–72", width 48–84"
- Zones 4–8

'Crimson Velvet' has a somewhat more relaxed architecture than 'Crimson Pygmy' and is much larger. Its foliage emerges crimson, maturing to sultry smoky-purple in summer. In fall, the hue changes again, this time to deep red. It is equally beautiful in all three stages. Like others of this genus and species, 'Crimson Velvet' produces small yellow flowers along the underside of its branches in spring. They are followed, in this cultivar, by showy scarlet drupes. The small oval to round leaves reach all the way to the ground, so it is attractive as a focal point or at the front of a large shrub border. To me, though, it really shines as a bold accent in a mixed perennial garden, where you can capitalize on its beautiful foliage colors by creating arresting combinations.

▲ 'Crimson Velvet' makes a great focal point in a mixed border.

Berberis thunbergii 'Orange Rocket'
Orange Rocket Barberry

- Full sun to part shade (fruiting and fall color are better in full sun)
- Tolerant of most well-drained soils, regardless of pH or fertility
- Foliage attractive all season, flowers in spring
- Height 54", width 24"
- Zones 4–8

This is a beautiful vertical shrub shaped like a rocket ready to launch. In spring, the foliage emerges early as a pleasing coral, then turns more coppery orange and finally finishes the season in dazzling scarlet. If you want an exclamation point in your garden, this could be it. Its exceptional color just sizzles in combinations

▲ 'Orange Rocket' foliage changes color throughout the season, becoming a coppery-orange in mid-summer. Marilyn and Sam Beachy's garden; Gee Denton's design.

▲ In late summer to fall, 'Orange Rocket' turns scarlet. Homer Garden Club's garden.

▲ Like many barberries, 'Orange Rocket' leafs out very early in the season—a real advantage for those with short seasons. Homer Garden Club's garden.

featuring green, burgundy, blue, and purple. With yellow or tangerine, it is dazzling. Spring brings on small buttery-yellow flowers on the undersides of its stems. Unlike round-shaped barberry shrubs, the flowers of 'Orange Rocket' are noteworthy precisely because the stems are held upright, revealing the underside to the viewer. The flowers are followed by attractive scarlet berries. Butterflies seem particularly attracted to this barberry.

▲ *Berberis thunbergii* var. *atropurpurea* 'Rose Glow' in Sharon and Jerry Froeschle's garden; Sharon's design.

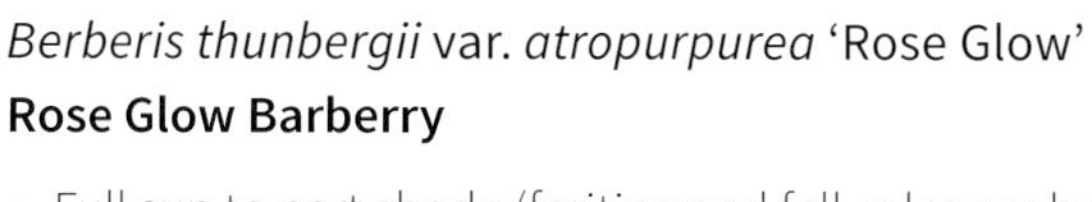

Berberis thunbergii var. *atropurpurea* 'Rose Glow'
Rose Glow Barberry

- Full sun to part shade (fruiting and fall color are better in full sun)
- Tolerant of most well-drained soils, regardless of pH or fertility
- Foliage attractive all season, flowers in spring
- Height 60–72", width 48–60"
- Zones 4–8

'Rose Glow' is a nice medium-sized, fast-growing, round-shaped shrub with the characteristic sharp thorns of its cousins. New growth appears on branches that arch out from a dense center. What distinguishes this variety is an interesting combination of colors as the growing season progresses. Early spring foliage is burgundy to purple and is followed by new shoots that are marbled with purple, rose, and splashes of cream. In this stage, 'Rose Glow' looks like a two-toned shrub, darker in the interior and lighter toward the surface—somewhat like sun-bleached hair. The effect is quite pretty. In the fall, the entire ensemble turns vibrant red. Although spring flowers are small and yellow, the bright red autumn berries that follow will lure your feathered friends into your late-season garden.

Cornus alba
Red Twig Dogwood

Redtwig dogwoods are vigorous, dependable and delightful, four-season deciduous shrubs. When taller than your winter snow level, their upright form will expose dark red stems that will sparkle vividly on a sunny winter day. Younger stems are redder than older ones. Though pruning is not necessary for the health and vigor of the plant, if you wish to increase the gleaming effect of your twigs, cut back 25 percent or so of the oldest stems to just above the crown in early spring each year. Moose will definitely sample these shrubs, so do protect them in winter; a burlap wrap or wire cage will accomplish that.

◄◄ *Cornus alba* 'Bailhalo' in Gail and Bob Ammerman's garden.

◄ 'Elegantissima' looks virtually the same as 'Bailhalo' but gets considerably larger. Vicki Rentmeester's garden.

Cornus alba 'Bailhalo' and *Cornus alba* 'Elegantissima'

Variegated Redtwig Dogwood

- Full sun to part shade, though stems have better color in full sun
- Humus-rich, well drained slightly acidic soil
- Attractive year round, blooms in spring to early summer
- 'Bailhalo': height 4', width 4'
- 'Elegantissima': height 10', width 10'
- Zones 3–7

'Bailhalo' and 'Elegantissima' dogwoods have lovely gray-green foliage with irregular ivory margins making them exceedingly versatile members of a mixed perennial garden. There is little difference between the two except size. 'Bailhalo' is a dwarf at three to four feet in height and width, while 'Elegantissima' will grow to ten feet in both directions. They are particularly stunning in a part shade location, where the pale margins really shine, lighting up their part of the garden. The stem color, however, will not be as intense as when sited in sun. In late spring to early summer, small white flower cymes reveal themselves but can be overshadowed by the foliage. The white berries that follow are definitely attractive additions to the overall impact of this plant. It will take some time for either to reach its maximum size, so purchase the largest specimen you can.

▲ The bright yellow, "painterly" leaf edges of *Cornus alba* 'Gouchaultii' create a sizzling effect. Quiet Creek Community.

Cornus alba 'Gouchaultii'

Tatarian Dogwood

- Full sun to part shade, though stems have better color in full sun
- Humus-rich, well-drained, slightly acidic soil
- Attractive year-round, blooms in spring to early summer
- Height 48–72", width 48–72"
- Zones 2–7

Though 'Gouchaultii' is similar to *Cornus alba* 'Elegantissima' and *Cornus alba* 'Bailhalo', its color combination is yellow and green rather than creamy white and gray-green. Because the variegation itself is bolder and the accent color is yellow, this selection makes a much more dramatic statement in the garden than the creamier options do. Each ovate leaf looks as if a creative painter tried a different pattern on it to see which he preferred. The overall effect is quite good-looking and bright. Small two- to three-inch-wide creamy spring flower clusters are followed by attractive blue-white drupes that attract birds. 'Gouchaultii' is a vigorous grower and will sucker somewhat, but all in all, it is a delightful addition to any mixed border. At Zone 2, it is also extremely hardy.

Philadelphus lewisii 'Blizzard'
Mockorange

- Full sun to part shade
- Moderately fertile, well-drained soil
- Blooms in mid-summer
- Height 60–84", width 60–72"
- Zones 3–8

When in bloom, *Philadelphus lewisii* 'Blizzard' steals the show. Brilliant, pure white, open-faced, usually five-petaled flowers form at the tip of every branch, including lateral ones, and, literally, cover the shrub. On top of their profusion, these pretty blossoms last several weeks and have prominent golden anthers that make them more interesting to see; they give off a divine orange orchard fragrance that is overwhelming. Enjoying a 'Blizzard' at its peak bloom is an awesome experience. Folks who are not enthused about

◀ 'Blizzard' is wonderfully fragrant.

▲ Each sweet-smelling flower has an open face with five petals and notable stamens.

this shrub complain that during the rest of the year, the simple, dark green, shiny, ovate foliage is uninspiring. To me, it is an excellent backdrop for colorful herbaceous perennials, but if you want the shrub itself to add another season of interest, then train an early or late clematis vine up into it. The clematis will seek the sunny side of your mockorange so that when the vine blooms, its flowers will adorn the ends of its host's branches. It's a great illusion and causes many people to wonder what the "mystery" shrub might be. Though many descriptions of 'Blizzard' indicate that it is compact, my experience is that is grows to at least six to seven feet in height, sometimes taller, and nearly that in width. The shrub is upright and somewhat vase-shaped, with long arching branches. It requires no pruning or special care of any sort, making it very low-maintenance.

Physocarpus opulifolius 'Dart's Gold'
Golden Ninebark

- Full sun to part shade
- Fertile, moist, well-drained acidic soil
- Blooms in spring, interesting all season
- Height 48–60", width 48–60"
- Zones 3–7

Physocarpus opulifolius, or ninebark, is best known for its popular purple-leaved cultivar 'Diablo'. Although 'Dart's Gold' shares the pleasing three-lobed, maple-like leaf shape of its close relative, it is much more assertive in both

▲ *Physocarpus opulifolius* 'Dart's Gold' turns lime green in part shade but brilliant chartreuse in full sun.

▲ As its foliage develops tinges of bronze in the fall, the burnished copper stems of 'Dart's Gold' become more vibrant, too.

color and vigor, especially in cold climates. 'Dart's Gold' will leaf out considerably earlier than the darker cultivars and will attain mature size sooner. Considering the shortness of the growing season in many cold-climate gardens, this earlier awakening is a distinct advantage. A little afternoon shade is beneficial in locales that have particularly hot summers.

Bright golden to chartreuse foliage makes 'Dart's Gold' prominent in part shade and a gleaming centerpiece in sun. During the spring, dense corymbs of small, pretty, white flowers form just above the foliage. This shrub has a mounded, somewhat cascading form. Interesting dark peeling bark provides a strong contrast to the bright foliage, though it is more prominent after the leaves have fallen. Before they fall, the leaves are infused with an attractive bronze hue. 'Dart's Gold' will perform better in moist, well-drained soil but it will also tolerate clay, dry, or rocky soil. It spreads slowly by suckers. Protect it from winter-browsing moose.

Potentilla fruticosa 'Mango Tango' (synonym *Potentilla fruticosa* 'UMan')
Mango Tango Cinquefoil

- Full sun
- Widely adaptable to different soils
- Blooms all summer
- Height 24–36", width 24–36"
- Zones 2–7

Though passed over by many gardeners who associate *Potentilla fruticosa* with the bedraggled, sulfurous yellow shrubs seen too often in harsh

▼ Super-hardy and incredibly tolerant of the worst possible conditions, 'Mango Tango' is, nonetheless, a real beauty. Stream Hill Park entry garden.

▶ *Potentilla fruticosa* 'Pink Beauty'. Homer Public Library's garden.

environments such as parking lots and along roadways in front of strip malls, the fact that these neglected shrubs survive at all in those locations and then bloom on top of it is testament to their tough-as-nails constitution. Happily, this genus has some exceptional representatives in a rainbow of colors that deserve consideration for a welcoming, low maintenance garden. One example is 'Mango Tango'. It has cheerful bicolored flowers of mango and burnt sienna. In bright sun, the mango appears more golden, making the blossoms glow softly. Rust-colored stems add an attractive accent, and tiny, glossy green leaves provide a delicate and fine texture useful as a contrast to larger leaved companions. This variety begins to bloom early in summer and continues until frost, attracting butterflies all along the way. For a compact, rounded, totally self-sufficient shrub, this one is hard to beat. Two other pretty cultivars are soft pink 'Pink Beauty' and creamy yellow 'Katherine Dykes'.

▲ Creamy-yellow *Potentilla fruticosa* 'Katherine Dykes'.

Salix purpurea 'Nana' (synonym *Salix purpurea* 'Gracilis')
Dwarf Arctic Willow, Dwarf Purple Osier

- Full sun to part shade
- Moist to wet soil of average fertility
- Attractive year-round
- Height 48–72", width 48–60"
- Zones 2–7

Salix purpurea 'Nana' is a wonderful cold-climate shrub that sways sinuously in a breeze. Newly emerging foliage is flushed with purple, becoming

◀ Because this shrub has been regularly "pruned" by moose, it hasn't achieved its potential size, but still retains a nice rounded shape. Quiet Creek Community.

▼ Dwarf arctic willow is an understated shrub with subtle beauty that is best viewed from nearby.

blue-green at maturity with silvery undersides. Flexible, thin, chocolate stems stand out against the attractive slender leaves. In fall, the leaves turn yellow making the stems even more eye-catching. Their flexibility makes the stems of this plant useful for weaving.

Like most willows, dwarf arctic willow does best in moist areas and can be grown in rain gardens, but it does well in average garden soil so long as it doesn't dry out. I've read that it forms male and female catkins in very early spring, but I've not observed them. Perhaps I wasn't looking at the right time. The stems are supposed to be distasteful to most animals, but that doesn't seem to bother moose. Fortunately, this shrub responds well to pruning, whether by the resident gardener or by a large hoofed animal.

Sorbaria sorbifolia 'Sem'
False Spirea Sem, Ash Leaf Spirea Sem

- Full sun to part shade
- Tolerant of most soil types and pH ranges, does best in average to moist soils
- Attractive all season, blooms in summer
- Height 36–48", width 42–60"
- Zones 2–7

▲ In cool weather—both spring and fall—'Sem' sports an overlay of apricot, a color that complements nearly every other hue.

This incredibly interesting and quite stunning shrub is often used in Britain but is barely known or grown in North America. Here, perhaps, this much improved cultivar suffers from association with the basic species, *Sorbaria sorbifolia.* That one has definitely earned its reputation as a real management headache because of its vigorous suckering nature. 'Sem' is much more compact and refined in its habits, and it has incredible foliage that changes from one beautiful color to the next. It may sucker a bit, especially when grown in highly fertile soil, but the suckers are thin and are easily clipped back.

This garden jewel is one of the first shrubs to leaf out in spring, heralding the start of the season as the crocuses come into bloom. New foliage begins as spiky reddish clusters reminiscent of witch hazel (*Hamamelis*) blossoms. As the foliage unfurls, its mountain ash–like leaves are an incredible apricot. This hue is joined by and blends with chartreuse as the weather warms. Ultimately, chartreuse becomes the dominant color, with a touch of green in the mix. Near the same time, the plant puts forth astilbe-like white flower plumes, making a very pretty combination. As the weather cools toward fall, apricot begins to reappear at the tips of the branches and ultimately becomes the dominant color. This uncommon and changing color palette provides a cornucopia of options in our combinations, some of which can be seen in the adjoining photos.

Sorbaria sorbifolia 'Sem' is a mounded, medium-sized shrub that holds its branches in horizontal layers, with new growth being a bit more upright. The foliage grows all the way to the ground, so there are no unattractive bare stems at the bottom. Because this shrub is so very hardy, gardeners in Zone 4 and higher should be able to successfully grow it in a container, thus having the beauty without any worry about suckering. Just be sure to get the cultivar 'Sem', not the species!

◂ Emerging foliage of *Sorbaria sorbifolia* 'Sem' is reminiscent of the late winter flowers of witch hazel.

▴ In mid-summer's warmth, the attractive lacy foliage exhibits a mix of chartreuse and green.

Spiraea
Spirea

A delightful, easy-to-grow, hardy group of shrubs, spireas are available in some spectacular cultivars that bloom profusely and change to a kaleidoscope of color in the fall. Most are of medium size and are somewhat rounded in silhouette, though several lovely varieties are gracefully arching. Although there are much larger spireas available, all those featured here mature at three to four feet in height and three to four feet in width. A bonus for gardeners who contend with moose and deer in their gardens is the distinct lack of interest these animals have with spireas. All the selections that follow are very attractive to butterflies—another bonus.

Spiraea x *bumalda* 'Denistar'
First Editions Superstar Spirea

- Full sun
- Average, medium, well-drained soil
- Attractive all season, blooms in summer
- Height 36–48", width 36–48"
- Zones 3–8

▴ New growth in warm-red complements more mature mid-green foliage and foretells the flower color that will soon add to the beauty of *Spiraea* x *bumalda* 'Denistar'. Gari and Len Sisk's garden.

▶ As the weather turns cold, the brilliant bronze leaves of 'Denistar' glow in the fall sunshine. Gari and Len Sisk's garden.

▶▶ For weeks on end, large flower clusters cover this pretty spirea, marketed as both 'Wilma' and 'Pink Parasols'.

A three-season performer, 'Denistar' displays deep red new-growth foliage. As its refined, pointed, and serrated leaves mature, they turn dark green, producing an exceptional contrast to the scarlet newer leaves. This striking combination continues throughout the summer until first frost, at which point the entire shrub will glow a brilliant copper. During mid-summer, the foliage show is enhanced by deep rose flowers arranged in medium-sized clusters at the ends of the branches. This Superstar has multiple stems that form a somewhat rounded silhouette. Being a naturally occurring dwarf, 'Denistar' requires no pruning to keep its size in bounds. It is another easy-care shrub that tolerates most soils as well as urban pollution and that is generally unappetizing to deer and moose.

Spiraea fritschiana 'Pink Parasols' (synonym *Spiraea fritschiana* 'Wilma') **Pink Parasols Spirea**

- Full sun
- Average, medium, well-drained soil
- Foliage attractive all season, blooms early to mid-summer
- Height 24–36", width 24–48"
- Zones 3–8

▶ Though the effect from a distance is solid pink, a closer look reveals deep pink flower centers. The pretty oval leaves of 'Pink Parasols' are considerably larger than those of many other popular spireas.

▶▶ Regardless of your favorite fall color, as the weather cools, you are likely to find it in the lovely potpourri of colors exhibited by *Spiraea fritschiana* 'Pink Parasols'.

This is an exceptional cultivar with particularly large three- to five-inch-wide, slightly rounded clusters of tiny, light pink flowers appropriately called 'Pink Parasols'. Each individual flower has a slightly darker pink center, adding a subtle two-toned aspect. The shrub is nearly covered with blooms for weeks from early to mid-summer, and even longer with dead-heading. These lovely flowers are framed by nicely contrasting blue-green, oval, and finely toothed foliage. Early in the season, as the leaves first open, they are flushed with a red tint, but the best part of the foliage show begins with the first signs of fall. It is then that the leaves turn a spectacular combination of orange, rose, and gold, with a touch of red or burgundy on colorful dark bronze stems. Stunning!

'Pink Parasols' is an upright, mounded, bushy shrub that will grow to three feet tall and three to four feet wide, making it a good fit for even small gardens. It tolerates a wide range of soils and is very easy to grow. If you choose to prune it at all, do so in late winter or very early spring before new flower buds form.

▲ Even the stems of *Spiraea fritschiana* 'Pink Parasols' turn brilliant copper in the fall.

Spiraea nipponica 'Snowmound'
Snowmound Spirea

- Full sun
- Average, medium, well-drained soil
- Blooms spring and early summer
- Height 24–36", width 24–48"
- Zones 3–8

For those of you who find bridal wreath spirea irresistible but who don't have room for a shrub that grows nine feet tall and up to eight feet wide, snowmound

◀ *Spiraea nipponica* 'Snowmound' has a delightful arching form. Homer Garden Club's garden.

▼ Shiny, dark green leaves are nearly covered by the profusion of its white flowers. Gari and Len Sisk's garden.

spirea will deliver a very similar look in a much more manageable size. Maxing out at three feet tall by four feet wide, this pretty shrub has the same delightfully arching habit. In spring and early summer its gently curving branches are covered on the upper side with showy corymbs of pure white tiny flowers. The effect is quite impressive. Small, shiny, very dark green leaves are hidden by all the flowers but add substantially to the attractiveness of this undemanding plant once the bloom has finished. It is a deserving recipient of the Royal Horticulture Society of Great Britain Award of Garden Merit.

Viburnum trilobum 'Bailey Compact'
Compact American Highbush Cranberry, Bailey Compact Cranberry

- Full sun to part shade
- Slightly acidic to neutral, moist, well-drained soil, but will tolerate poorly drained soil
- Blooms spring to early summer, fall color and berries
- Height 48–74", width 36–52"
- Zones 2–7

In the spring, dainty white lace-cap flower clusters open with small, fertile flowers in the center, surrounded by much showier sterile ones. They're a handsome and satisfying addition to the spring garden. The bold shape and shiny texture of this selection's rich green foliage as well as its dense, upright nature, makes it an excellent backdrop for summer-flowering plants. When the chill of fall descends on the garden, however, 'Bailey Compact' steps into the limelight and puts on a fabulous show of brilliant red and burgundy hues both in its leaves and in its pendant berry clusters. The berries will often persist well into winter. Though somewhat tart, the fruit is edible and often used in making preserves and chutney.

This shrub must be protected from moose in winter but is otherwise very tolerant of a broad range of growing conditions. It needs no pruning if sited where it has room to grow to its natural size and shape.

◂ Beautiful foliage is the hallmark of *Viburnum trilobum*. The cultivar 'Bailey Compact' delivers this exceptional look in a more diminutive size.

▾ Fall temperatures trigger the brilliant crimson color so beloved by gardeners. Bear Creek Winery Gardens, Homer, Alaska.

TREES

he most statuesque members of a mixed garden are, of course, its trees. Their relative size alone puts them into the star category, whereas their other attributes provide structure, mass, vertical dimension, spring or summer flowers, and often brilliant fall color—plus berries, drupes, or cones. Additionally, the bark of deciduous trees adds year-round color and texture to the garden tableau. The architecture of winter's leafless branches instills a sense of drama to what can be an otherwise uninspiring snow-covered landscape, while evergreens provide year-round color and texture. For folks with small gardens, there are also some delightful small trees that can be successfully incorporated.

Most trees are long-lived, expensive, and difficult to move, so determining their location in the garden before planting them is something that deserves careful thought and consideration. In locations with moose or hares, it is necessary to protect them from being permanently disfigured or even killed by these wild, unruly, and ravenous creatures.

Acer ginnala
(synonym *Acer tataricum subsp. ginnala*)
Amur Maple

- Full sun to part shade
- Best in moist, well-drained soil, tolerates a range of pH levels and occasional drought
- Spring flowers, fall foliage
- Height 10–20′, width 12–20′
- Zones 2–8

Anyone who knows Amur maples values their incredibly beautiful fall foliage colors. As temperatures drop, the glossy green leaves first take on a bit of yellow and orange; the colder it gets, the more the hues intensify, often to flaming red. Leaves are about three inches long, generally have three lobes with slightly irregular edges, and are attached to the branches with pink

◀ *Acer ginnala* really shines as the weather cools. Rita Jo and Leroy Shoultz's garden; Rita Jo's design.

▲ Fabulous fall foliage is worth the wait.

petioles. But there is much more to this recipient of the Royal Horticulture Society's Award of Garden Merit than its glorious foliage.

In spring, this small tree is adorned by fragrant, pendulant, creamy-white flowers arranged in small clusters. These are eventually replaced by winged seedpods that are called samaras that also turn red as they mature. Although the bark is an unexceptional gray to brown, *Acer ginnala* can be pruned into enticing ornamental configurations that are best revealed if the tree is "limbed" up somewhat. In fact, it is even used in Japan for bonsai.

This hardiest of all the maples does best in cool summers. Its small size makes it an excellent choice for smaller landscapes. Because of its tolerance for different soil types and moisture levels, it is often used in stressful urban environments as well. As a young tree, its natural silhouette is vase-shaped, developing a rounded crown with maturity. Most commonly sold in multi-stem configurations, it can sometimes be found as a single-trunk specimen. One word of caution: it has been found moving into wild areas in parts of the upper midwestern United States. Please check with local authorities for updated status if you garden in this area of the country.

Acer platanoides 'Royal Red'
Royal Red Maple, Red Norway Maple

- Full sun
- Best in average to moist conditions, tolerant of various soil types and pH levels
- Flowers in early spring, foliage all season
- Height 50′, width 40′
- Zones 3–7

Considered the hardiest of the *red* maples, this Norway giant is at the other end of the size spectrum from petite *Acer ginnala*, but that's not where the contrasts end. *Acer platanoides* 'Royal Red' begins the season with lemon-yellow flowers along its leafless branches, which are followed by shiny, leathery-looking, five inch wide, deep burgundy foliage. These stunning leaves have five lobes that are pointed at the tips, reflecting what I've always thought of as classic maple-shaped leaves—perhaps because I grew up in Philadelphia, where the streets were lined with hundred-year-old Norway maples (*Acer platanoides*). Although a relatively new cultivar, fast-growing 'Royal Red' is also purported to have a hundred-year life expectancy, making it the ideal specimen to plant as a hereditary legacy. Foliage stays burgundy throughout the summer, turning somewhat deeper purple in fall. If removed from the tree, the burgundy petioles will weep white sap.

◂ *Acer platanoides* 'Royal Red' makes an excellent focal point and, when it matures, a wonderful shade tree. Joan Splinter and Don Felton's garden; Joan's design.

▾ Shiny, leathery, five-lobed, burgundy leaves in the classic maple shape are at least five inches across. Stems and petioles are also burgundy. Joan Splinter and Don Felton's garden.

The trunk is perfectly straight and rises approximately seven feet before the canopy branches out, so this is a good candidate as a shade tree. The canopy is somewhat conical in younger specimens but becomes rounded as it matures. Though it is slow-growing, the ultimate height and width of 'Royal Red' makes careful placement in the garden critical.

Like many maples, this lovely tree is tolerant of city environments, including pollution and is notably disease-free. It's a wonderful choice, especially if you like dark foliage.

Betula papyrifera
Paper Birch, White Birch, Canoe Birch

- Full sun to part shade
- Slightly acidic, moist loamy soil
- Bark attractive year round, good fall color
- Height 50–60′, width 25–35′
- Zones 2–7

▲ *Betula papyrifera* naturally grows in large stands such as the one seen here at the Alaska Botanical Garden.

▲ The most notable characteristic of this tree is its magnificent bark, which gets whiter as the tree matures.

▲ A large paper birch can also serve as a hiding spot for a curious porcupine!

Noted for its luminous white bark, paper birch is an excellent tall and extremely hardy tree. It is native to Alaska and Canada and does best in areas with dependable winter snow cover and cool summers. Although it can be used as an individual tree that serves as a dramatic focal point, it has a much greater impact when planted in a small group or even a large grove. As the tree matures, its crown gets more rounded and the pure white bark exfoliates in strips, revealing the caramel-colored bark beneath. Combined with scattered charcoal markings, this range of colors is both subtle and dazzling at the same time and provides the gardener with many options for creating exciting combinations.

The irregularly toothed leaves are an attractive shape usually described as ovate but are actually broader at the base and taper to a point. As the tree buds out in spring, the foliage is so light green as to be nearly chartreuse. It's a wonderful contrast with dark-needled conifers such as spruce. Leaves open green and then become a warm, buttery yellow in the fall. Male and female flowers appear on the same tree but are barely noticeable.

Crataegus x *mordenensis* 'Toba'
Toba Hawthorn

- Full sun to part shade
- Well-drained soil, tolerant of salt as well as of different soil types and pH levels
- Blooms early spring, berries in fall
- Height 15–20′, width 10–15′
- Zones 3–7

Fragrant, double white flowers with a hint of pink sit atop the branches of this perfectly lovely small tree. As the flowers age, they change to pink, creating a two-toned impression along the way. Just as these early flowers attract bees, so the bright red round berries that follow in autumn are a magnet for songbirds. What a delightful dual purpose for those wishing to attract the birds and bees to their gardens!

Introduced in Manitoba, Canada, where hardiness zones range from Zone 4 at the U.S. border to Zone 2 in the north of the province, *Crataegus* x *mordenensis* 'Toba' is considered one of the hardiest hawthorns. Its dark green, two- to four-lobed, serrated leaves are very appealing and shiny, particularly on a sunny day. They turn buttery yellow in fall.

The bark is somewhat rough and appears gray-green or golden-green, depending on light conditions. While many descriptions of this cultivar describe the trunk as twisted and the branches as thorny, I have not seen either characteristic on the specimens I've planted or observed in the gardens

◂◂ This is a perfect choice for a small garden. Fragrant double flowers open white but fade to pink as they age. Joan Splinter and Don Felton's garden; Joan's design.

◂ Luminous two- to four-lobed leaves add to the charm of this selection. Joan Splinter and Don Felton's garden.

of others. Perhaps these traits appear only in much more mature trees than those I've known. The silhouette begins as somewhat vase-shaped, maturing slowly to a more or less rounded form. Because the leaves will become tattered in windy locations, be sure to select a spot that is sheltered. Though tolerant of many soil types and pH levels, 'Toba' will succumb to standing water.

Hippophae rhamnoides
Seabuckthorne, Sea Berry

- Full sun
- Moist, well-drained, sandy, even salty soil with neutral to alkaline pH
- Flowers in spring, fruit in late summer to fall on female plants
- Height 12–20′, width 20–30′
- Zones 3–8

If you've ever marveled at the gray-green beauty and twisted branches of an olive tree (*Olea europaea*) while traveling in warm, Mediterranean climates and wished you could grow one, this is the answer to that desire. Very hardy and undemanding, seabuckthorne sports the same kind of narrow, gray-green leaves, though at three inches, they are longer and more feathery than those of the European classic.

▶ Though often described as a shrub, Sea Buckthorne can be pruned into a very interesting specimen tree, as this one has been in Denice and Roger Clyne's garden.

Rather than developing olives, which are a messy nuisance in landscape applications, this selection offers small yellow-green flowers prior to leafing out, ultimately followed by three-eighths-inch-wide, bright orange, Vitamin C–rich, edible berries. Although most references indicate that both a male and female are required for fruit production—and perhaps that is true for agricultural purposes—my experience has been that a single tree will produce some fruit in late summer to early fall. It can be used to make jam, juice, or even wine.

▲ Edible, bright orange berries are tart, but extremely high in vitamin C.

Rough, fissured, gray-brown bark covers the trunk and twisted branches of this architecturally interesting, multi-stemmed specimen tree. Unpruned, it can spread up to forty feet in width. If you garden where heavy snowfall is to be expected, I recommend guiding its growth in a slightly more upright configuration. View this as an opportunity for fun, because pruning *Hippophae rhamnoides* a little bit each year over an extended period can result in a very artistic, sculptural, focal point not unlike a full-sized bonsai.

Native to coastal sand dunes and screes, as well as mountain river banks, seabuckthorne is tolerant of salty soils. In fact, it performs best in infertile soil. It does sucker somewhat, making it a good candidate for soil stabilization applications. Interestingly, though somewhat rare in the United States, this tree is prevalent throughout Europe, including Finland, where it migrated from Nepal after the last ice age. How's that for a cocktail party tidbit?

Malus 'Royalty'
Flowering Crab Apple

- Full sun
- Moist well-drained soil
- Blooms in spring, foliage is interesting all season
- Height 12–16′, width 8–14′
- Zones 2–6

▲ *Malus* 'Royalty' has dark purple foliage and deep rose-colored flowers in spring.

Just one of the myriad of delightful flowering crabapples available to cold-season gardeners, 'Royalty' has fabulously glossy deep purple foliage. In mid-summer, its dark, oval leaves exhibit a minor flush of green and then turn a brilliant dark red in the crisp temperatures of fall. As if that's not enough, 'Royalty' produces pretty, single, deep rose flowers in spring, followed by small edible red fruit. The mahogany bark continues the warm color scheme.

There are so many other notable flowering "crabs" that it can be difficult to choose your favorite. *Malus* 'Dolgo' has pink buds that open to pure white flowers, which are followed by red-purple fruit. *Malus* 'Makamik' has rosy red blossoms and bronze-infused green foliage that turns bright yellow in fall while *Malus* 'Thundercloud' sports dark pink flowers and purple leaves that turn a nice orange-yellow after frost. Both produce small, round red fruit.

Picea glauca 'Conica'
Dwarf Alberta Spruce

- Full sun to part shade
- Fertile, humus-rich, moist, well-drained soil
- Attractive year round
- Height 10′, width 4–5′
- Zones 2–7

Without any pruning, this evergreen is perfectly conical in shape—a very interesting and useful architectural configuration. Its dense, dark green needles are so soft that you will be compelled to reach out and touch them. Furthermore, the fine texture of dwarf Alberta spruce accentuates and offers a unique contrast to broad-leafed plants. An evergreen that is virtually

▼ *Picea glauca* 'Conica' in a simple combination of texture extremes. It does best in full sun but will perform just fine in a part shade location.

▶ 'Conica' retains its perfect conical shape without pruning. Quiet Creek Community.

maintenance-free, it requires no care at all except an assurance of adequate moisture and planting in a location that will reliably meet its cultural needs.

Picea glauca 'Conica' is a very popular and well-known conifer in the United States. Obviously, many folks have recognized it as a cool plant. A similar evergreen, *Picea glauca* 'Sander's Blue', has needles with a more intense blue tint and is hardy to Zone 3 rather than to Zone 2.

Populus tremuloides

Quaking Aspen, Poplar

- Full sun
- Best in fertile, humus-rich, moist, well-drained soil
- Bark attractive year round, good fall color
- Height 45–55′, width 20–30′
- Zones 1–6!

Did you notice the zone? Yes, this native of Alaska and Canada is truly hardy to Zone 1! It does best in locations that have cold winters and a cool growing

◀ When the weather becomes crisp and cold, the trembling leaves of *Populus tremuloides* turn brilliant golden-yellow.

▼ Not only does quaking aspen entice us with its fluttering foliage, but its bark is also appealing in subtle greenish hues with raised horizontal markings.

season, so it is not recommended for those regions having consistently hot, humid summers. Quaking aspen takes its common name from the fact that the slightest breeze will cause its nearly round, finely toothed, dark green leaves to flutter merrily. Because the underside of the foliage is slightly lighter, being almost silvery-green, it creates the distinct impression that the leaves are changing colors. During fall, when the foliage turns brilliant yellow, the trembling effect is totally mesmerizing. This is a tall tree with a somewhat pyramidal shape; it is especially effective in groupings. The bark of *Populus tremuloides* varies from greenish gray to tan, even white. My strong preference is for those with a decidedly green tint to their bark, but you may find the other options more appealing. Raised, horizontal markings add welcome complexity to the appearance of the bark.

A word of caution: In the wild, quaking aspen spreads by suckers that are clones of the parent plant. This natural tendency of the plant also occurs in cultivation, so give it plenty of room, and keep it safely away from structures and foundations. This trait, and the tree's rapid growth, makes quaking aspen an excellent choice for revegetation projects.

Prunus maackii
Amur Chokecherry

- Full sun to part shade
- Average, moist, well-drained soil
- Spring flowers, brilliant bark all year
- Height 25–35′, width 20–25′
- Zones 2–6

If you are drawn to interesting, colorful bark, then you ought to consider *Prunus maackii.* Its trunk and branches are covered with glossy, cinnamon-colored bark that is adorned with slightly raised, horizontal, light-colored markings. The combination is beautiful in summer but is absolutely brilliant in winter after fallen leaves reveal the tree's entire broadly rounded structure. Before it reaches maturity, the bark will often exfoliate in strips, adding even more texture to the display.

Clusters of small white flowers bloom in spring and are followed by the emergence of small, dark red, round fruit. Foliage is green, slightly toothed, and oval to elliptical in shape. It gradually turns pale yellow-green, with leaves dropping early in fall, revealing the true glory of this interesting specimen tree.

◄ If you enjoy interesting bark, consider *Prunus maackii* in a mixed planting where its unique coloration can be used to create spectacular combinations like this one in Kathy and Mike Pate's garden.

▲ The absence of leaves during winter focuses our attention on its graceful form and brilliant bark. Georgeson Botanical Garden, Fairbanks, Alaska.

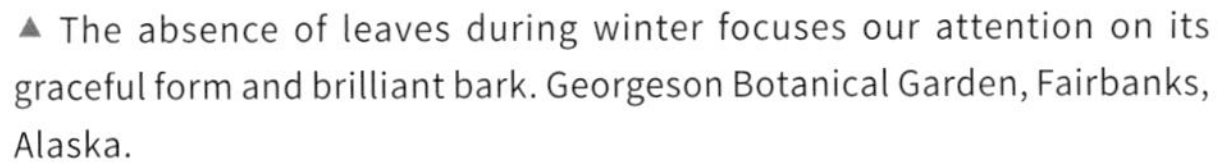

▲ Close-up of a mature specimen of *Prunus maackii* on the campus of the University of Alaska Anchorage.

Prunus virginiana 'Bailey Select'
Bailey Select Schubert Cherry

- Full sun
- Average to medium fertile, moist, well-drained, acidic to alkaline soil
- White flowers in spring, attractive foliage all season
- Height 20–30′, width 15–20′
- Zones 2–7

One of the things that sets midsized, cold-hardy 'Bailey Select' apart from other burgundy-leaved options is its chameleon-like character. It begins the growing season cloaked in green foliage, with ample clusters, of ornate, fragrant, white flowers hanging below the green-clad limbs. By mid-June, the oval, three- to four-inch-long foliage begins a transformation, turning a red-infused dark burgundy. During this transition, the tree has a fascinating two-toned appearance but ultimately becomes solidly dark-leaved. After fall frosts, the leaves turn an outstanding red but also begin to wilt.

After the flowers fade, clusters of sour black berries form on this self-pollinating selection. Though much beloved by birds, the edible berries are best eaten by humans in jams or other dishes to which sweeteners are added. Because birds will flock to these trees as berries ripen, you are best advised to plant them well away from patios and walkways to avoid the inevitable droppings.

The architecture of *Prunus virginiana* 'Bailey Select' is tall and oval, with the canopy beginning from four to eight feet above the ground. The bark is slightly ridged and distinctly gray, making a pleasing combination with burgundy. I personally like to prune up this tree to reveal more of the bark because its color complements the foliage so well; however, pruning is not necessary. Though its growth rate is usually described as medium, in severely cold climates 'Bailey Select' grows faster and more vigorously than many others. It is truly a cold-climate tree, not performing well in hot areas. Its cultivar name honors Bailey Nursery of St. Paul, Minnesota, where it was introduced. Though tolerant of urban pollution, this tree will succumb to standing water. Eastern tent caterpillars find it a hospitable home.

▶ In June, foliage turns burgundy, creating a two-toned effect while in transition. Cold weather at the end of the season changes the foliage once again, this time to bright red. Though they are normally seen with a single trunk for the first four feet, this example branches out quite close to the ground. Kathy and Mike Pate's garden.

▼ In spring, *Prunus virginiana* 'Bailey Select' is covered in clusters of large, fragrant, white flowers amid green foliage. Rita Jo and Leroy Shoultz's garden and design. Photo courtesy of Rita Jo Shoultz.

▶ Gun-metal gray bark offers a nice complement to the various foliage costumes 'Bailey Select' dons throughout the year. Kathy and Mike Pate's garden.

VINES

Vines can be annual or perennial, herbaceous or woody. When supported by a trellis or other structure, they offer a charming and often colorful vertical element. This is especially effective in gardens that are too small to accommodate trees. Allowed to ramble on the ground, vines can add a bit of comedic whimsy to your show. You can train a small vine into a shrub, adding another season of bloom to its host. They can also be useful in creating a privacy screen or an attractive backdrop for other members of your garden's cast.

Annual vines will bloom throughout one growing season, adding lots of flower color (see Section V, "Annuals"). Perennial vines, on the other hand, can be an enduring feature in your garden in much the same way that a tree or shrub can. Their diverse foliage, fall color, seed heads, stems, and flowers will all add beauty, interest, and allure to your garden. This section focuses on perennial vines.

◄ An intriguing 'Red Wine' climbing monkshood whose seeds are available via several online seed suppliers. My plant came from fellow gardener Teena Garay.

▼ There is also a pretty lavender-blue selection of climbing monkshood.

Aconitum aff. hemsleyanum and *Aconitum aff. hemsleyanum* 'Red Wine' **Climbing Monkshood**

- Part sun, tolerates full sun
- Cool, moist, fertile soil, but will tolerate most soils and a broad range of pH
- Blooms mid-summer to early autumn
- Height 6–10′, width 1–3′
- Zones 5–8

An intriguing, twining herbaceous climber that grows to up to ten feet in a season and that has attractive medium to dark green, deeply cut, five-lobed leaves, vining monkshood hails from the mountains of China. To quote the incredibly knowledgeable plantsman and founder of Heronswood Nursery, Dan Hinkley, "Trying to put an actual name on a climbing *Aconitum* from Asia is tricky work." Hence, at his recommendation, I've used the designation *Aconitum aff. hemsleyanum*. The abbreviation "*aff.*" stands for the

Latin word *affinis,* meaning related to or similar to. Dan also counsels that there are numerous good and hardy climbing monkshoods from China and other areas in Asia, including *Aconitum bulbifera* and *Aconitum episcopale.* The take-away from all this technical detail is that if you find a vining monkshood, regardless of what it's labeled, buy it and try it. It's a quintessentially "cool" plant.

In their native habitat, these plants have somewhat variable color, but the most commonly seen color is a soft lavender-blue. An exceptional strain sold as 'Red Wine' is the color of an expensive California burgundy wine. For those of us who love the jewel tones, both of these are fantastic. I grow one on either side of a large rusted-metal arbor and let them intertwine at the top. They are also an excellent choice for training onto shrubs, because the plant is airy and light in weight. Flowers open in mid-summer and continue until frost; bees love them. A note of caution: as with nonclimbing monkshoods, all parts of the plant are toxic.

Actinidia kolomikta 'Arctic Beauty'

Hardy Kiwi Vine, Arctic Kiwi

- Full sun to part shade
- Loamy, medium moisture, well-drained, acidic to neutral soil
- Foliage all season, small flowers in spring
- Height 20–25′, width 6–10′
- Zones 4–8

▲ If you decide to grow hardy kiwi vine, be sure to give it an incredibly sturdy structure on which to climb. Gabriela Hussman and Conrad Schaad's garden and design.

▶ Male foliage of this kiwi is much more colorful than that of the female. Growing the two together provides beauty *and* fruit production. Gabriela Hussman and Conrad Schaad's garden.

'Arctic Beauty' is an exuberant vine that needs a *very* sturdy structure on which to grow, for it readily attains a height of twenty feet or more. There are both male and female plants, easily identified by their distinct colorations; both are necessary for fruit production. Fruit is small and smooth-skinned but has a yummy taste similar to that of the bigger fuzzy ones found at the supermarket.

Much like the males of the avian world, for ornamental purposes, the male is preferred because it has an unusual, three-colored, intensely variegated foliage, whereas the female's leaves are simply green with an occasional touch of color. Grown together, though, you get the best of both—fruit *and* beautiful foliage. Male kiwi foliage begins the season as green, then slowly changes to white, bright pink, and green in painterly and varied combinations. Each leaf is somewhat different than the next. Though the vines are very hardy, early-season growth can be damaged by a late frost. Small white spring flowers are usually hidden by the pretty, heart-shaped foliage. Fortunately, they are deer-resistant.

Clematis
Clematis, Russian Virgin's Bower, Old Man's Beard

Sometimes called the queen of the vines, clematis is available in a vast range of sizes and bloom types. Each has its own unique charm, though the pruning regimens can be confusing and somewhat unsettling. Typically, if the species blooms early in the season, its flowers open on stems from the prior year—so prune after they bloom. Vines that bloom late in the season are flowering on current-year vines and so can be trimmed in early spring. My approach is to site a vine where it will have plenty of room to grow and then not bother with pruning at all. You will often see the admonition to shade the roots of clematis so that they do not get too hot. Cold soils combined with cool summers make this unnecessary, but if you garden with hot summers, you'd be wise to follow this guidance.

Clematis alpina 'Constance'
Alpine Clematis

- Full sun, bright shade
- Fertile, humus-rich, well-drained soil
- Flowers in spring to early summer, seed heads in late summer to fall
- Height 6–10', width 5'
- Zones 3–9

While the showy, enormous-flowered clematis are incredibly enticing, a much more reliable and charming choice for cold-climate gardeners comes in the form of the alpines. These hardy members of the vining clematis clan make up in quantity for their smaller blossom size. In spring and early summer, the vine is covered from top to bottom in open, nodding, bell-shaped flowers. 'Constance', a favorite, produces two- to three-inch-wide, semi-double, deep rose blossoms with just a hint of white on the petal edges and a prominent white center for weeks. This prolific bloomer will often produce a second and smaller flush of flowers in late

▲ The alpines are the hardiest and most dependable clematis for extreme cold-climate gardeners. Because they bloom very early in the season, you might enjoy a second bloom during warm summers. Gari and Len Sisk's garden.

▶ When the blossoms of *Clematis alpina* 'Constance' open fully, they reveal notable white stamens. Gari and Len Sisk's garden.

summer that intermingle sweetly with the eye-catching, fluffy seed heads from the first bloom. It's a three-month-long show. A similar cultivar, *Clematis alpina* 'Willy', sports pale pink flowers with creamy anthers. The alpines are an excellent choice for training into a shrub. Should pruning be necessary, give alpine clematis a light trim immediately after it blooms.

Clematis tangutica
Russian virgin's bower

- Full sun, bright shade
- Fertile, humus-rich, well-drained soil
- Flowers mid-summer to late autumn
- Height 15–20′, width 6–10′
- Zones 2–9

Clematis tangutica is big, bold, brash, vigorous, and long-blooming. Its bright yellow bell-shaped blossoms resemble little lanterns and appear in abundance from mid-summer until fall. As each flower fades, it is replaced by an equally delightful fluffy seed head. These seed heads engender the perfectly appropriate nickname "old man's beard." *Clematis tangutica* will grow to twenty feet in height and six to ten feet in width, so give it plenty of room and a very stout support upon which to climb. This clematis blooms on the current year's vines, so do any necessary pruning in early spring.

◀ Vigorous *Clematis tangutica* is very hardy and has bell-shaped yellow blossoms. During the summer, they are joined by wonderful fuzzy seed heads. It needs a substantial structure on which to grow.

GLOSSARY

Axillary shoot: A shoot in the angle between the main stem and a side stem or leaf.

Bare root: Said of a plant that has no soil on its roots. Used as a method of offering plants for sale, especially where shipping costs are high.

Berry: A soft-fleshed fruit that contains one or more seeds.

Central leader: The uppermost portion of the highest and stoutest major structural limb of a tree.

Crown: The growing point of a plant. It is usually just above or just below the soil surface.

Corymb: A flat or slightly domed cluster of flowers in which the outer flowers usually open first.

Cultivar: Short for "cultivated variety."

Cyme: Similar to a corymb, but the outer flowers open last.

Drupe: A stone fruit encased in fleshy material.

Farinose: White, mealy coating found naturally on the leaves, stems, or flowers of some plants.

Genus: A grouping of plants having similar characteristics. In binomial nomenclature (which gives us a plant's botanical name), the first word is the genus. It is italicized and capitalized.

Half-hardy perennial: A plant that survives only mild winters. It is usually grown in cold climates as an annual.

Inflorescences: A cluster of flowers on a single stem. Often used to describe the flowers of grasses.

Naturalizes: Multiplies and spreads over time. Often said of bulbs and other introduced plants that multiply by division and by seed.

Panicle: A compound, loose, and branched cluster of flowers.

Patented plant: A plant whose propagation and sale have been restricted by the owner of the patent. To obtain a plant patent, the inventor (breeder or discoverer) of a new plant must apply to the U.S. Patent and Trademark Office.

Plugs: Small seedlings usually grown in trays with multiple cells.

Pollinator: Anything that moves pollen from one flower to another. Wind, water, bees, butterflies, hummingbirds, and even mosquitos are pollinators.

Raceme: A cluster of flowers radiating from an unbranched stalk and attached to it by short pedicels.

Rhizome: Horizontal, underground, fleshy stems that are sometimes branched.

Root-bound: A condition in which the roots of a potted plant have exceeded the space in its pot and have begun to circle around, often tangling themselves beyond repair.

Scape: A flower stem with no leaves.

Sepal: A structure formed outside the petals of a flower. As a group, they're known as the calyx.

Species: A smaller grouping within a genus—a refinement of the genus such that the members of the smaller group have more in common with each other than they do with other species in the genus. The species is the second word in a botanical name. It is italicized and lower case. The abbreviation "sp." refers to a single species within a genus; "spp." indicates more than one species.

Sport: A naturally occurring mutation of a parent plant.

Stamen: The male part of a flower, necessary for reproduction.

Sucker: A root shoot that travels some distance underground from the parent plant.

Tender perennial: A perennial that cannot survive temperatures below 40 degrees Fahrenheit. It is even less tolerant of cold than a "half-hardy perennial" and is grown in cold climates as an annual.

Turgid: Swollen or full. Refers to a well-watered plant.

Umbel: A flat or rounded flower cluster with stems emanating from the same point.

Variety: A subset of a specific species that differs slightly from other members and whose characteristics will reproduce in future generations. This designation is also used when the members of a group of plants are nearly identical to the genus and species but have a notable difference, such as in foliage color. Abbreviated "var." in a plant name.

MATERIALS AND WORKS CONSULTED

Adams, Brenda. *There's a Moose in My Garden: Designing Gardens in Alaska and the Far North.* Fairbanks: University of Alaska Press, 2013.

American Primrose Society. www.americanprimrosesociety.org.

Armitage, Allan M. *Herbaceous Perennial Plants.* Athens: University of Georgia Press, 2008.

"Beneficial Insects 101." Planet Natural. www.planetnatural.com/beneficial-insects-101/.

Blue Stem Nursery. www.bluestem.ca.

Brent and Becky's 2016 Fall Planted 2017 Spring Flowering Catalogue. www.brentandbeckysbulbs.com.

Burns, Mary Jo. "Growing Primula in South-Central Alaska." Alaska Master Gardeners Anchorage. www.alaskamastergardeners.org/primulas_in_alaska.html.

Brickell, Christopher, and Judith D. Zuk, eds. *The American Horticultural Society A–Z Encyclopedia of Garden Plants.* New York: DK Publishing, 1997.

"Canada's Plant Hardiness Zones." Natural Resources Canada. Last updated 2014. www.planthardiness.gc.ca/images/PHZ_2014_CFS_Map_30M.pdf.

Cooperative Extension Service University of Alaska. *Sustainable Gardening.* Fairbanks: University of Alaska Fairbanks Press, 2010.

Darke, Rick. *The Encyclopedia of Grasses for Livable Landscapes.* Portland, OR: Timber Press, 2007.

Dirr, Michael A. *Dirr's Hardy Trees and Shrubs: An Illustrated Encyclopedia.* Portland, OR: Timber Press, 1997.

DiSabato-Aust, Tracy. *The Well-Tended Perennial Garden: Planting & Pruning Techniques.* Portland, OR: Timber Press, 1998.

Emden, Eva van. "Science Writing and Editing: How to Write Scientific Names." Last updated March 14, 2011. http://blog.vancouvereditor.com/2011/03/science-writing-and-editing-scientific.html.

Engebretson, Don, and Don Williamson. *Tree and Shrub Gardening for Minnesota and Wisconsin.* Edmonton, Canada: Lone Pine Publishing, 2005.

Geneve, Robert. "Flower Shapes." University of Kentucky College of Architecture, Department of Horticulture. www.uky.edu/Ag/Horticulture/Geneve/teaching/PLS%20220/Flowers/Flower%20shapes.pdf.

Haynes, Cindy. "Cultivar versus Variety." *Horticulture and Home Pest News*, Iowa University Extension and Outreach. Last updated February 6, 2008. www.ipm.iastate.edu/ipm/hortnews/2008/2-6/CultivarOrVariety.html.

The Herb Society of America. *The Herb Society of America Style Manual.* www.herbsociety.org/documents/styleman.pdf.

Lehrer, Jonah. "Smell and Memory." Science Blogs. Last modified November 9, 2009. http://scienceblogs.com/cortex/2009/11/09/smell-and-memory/.

Mathews, Brian, and Phillip Swindells. *The Complete Book of Bulbs, Corms, Tubers, and Rhizomes: A Step-by-Step Guide to Nature's Easiest and Most Rewarding Plants.* Pleasantville, NY: Reed International Books Limited, 1994.

McNeill, J., F. R. Barrie, W. R. Buck, V. Demoulin, W. Greuter, D. L. Hawksworth, P. S. Herendeen, S. Knapp, K. Marhold, J. Prado, W. F. Prud'Homme Van Reine, G. F. Smith, J. H. Wiersema, and N. J. Turland. *International Code of Nomenclature for Algae, Fungi, and Plants* (Melbourne Code). Königstein, Germany: Koeltz Scientific Books, 2012. www.iapt-taxon.org/nomen/main.php.

Missouri Botanical Garden. www.missouribotanicalgarden.org.

Paghat the Ratgirl. "*Saxifraga* x *arendsii* 'Triumph', a Reliable Stand-by." www.paghat.com/saxafraga.html.

Perennials.com. www.perennials.com.

"Plant Hardiness Zones of Canada." Agriculture and Agri-Food Canada. http://open.canada.ca/en/apps/plant-hardiness-zones-canada.

Phillips, Ellen, and C. Colston Burrell. *Rodale's Illustrated Encyclopedia of Perennials.* Emmaus, PA: Rodales Press, 1993.

Pratt, Verna E. *Wildflowers Along the Alaska Highway.* Anchorage, AK: Alaskakrafts, 1991.

Reinhardt, Didier, and Cris Kuhlemeier. *EMBO Reports* 3, no. 9 (2002): 846–851. www.ncbi.nlm.nih.gov/pmc/articles/PMC1084230/.

Suzuki, Wendy. "Making Sense of Scents: Why Odors Spark Memory." Live Science. Last modified on April 17, 2015. www.livescience.com/50525-why-odors-spark-memory-podcast.html.

University of California Division of Agriculture. http://ucanr.edu.

U.S. Department of Agriculture. "Title 190." In *National Plant Materials Manual.* www.nrcs.usda.gov/Internet/FSE_DOCUMENTS/stelprdb1042144.pdf.

Zuzek, Kathy, and Karyn Vidmar. "American Cranberrybush (*Viburnum trilobum*)." University of Minnesota Extension. www.extension.umn.edu/garden/yard-garden/trees-shrubs/viburnum-trilobum/.

INDEX

Italicized numbers indicate photographs.

Acer ginnala (Amur maple), *194,* 195–96
Acer platanoides 'Royal Red' (red Norway maple; royal red maple), 196–97, *197*
Acer tataricum subsp. ginnala (Amur maple), 195–96
See also *Acer ginnala* (Amur maple)
Achillea millefolium 'Paprika' (yarrow), *34*
Achillea millefolium 'Terracotta' (yarrow), *42*
Aconitum aff. hemsleyanum (climbing monkshood), 209–10
See also *Aconitum aff. hemsleyanum* 'Red Wine' (climbing monkshood)
Aconitum aff. hemsleyanum 'Red Wine' (climbing monkshood), *208,* 209–10
Aconitum 'Blue Lagoon,' 125
Aconitum bulbifera, 210
Aconitum carmichaelii 'Bicolor' (bicolor monkshood), 124–25
See also *Aconitum* x *cammarum* 'Bicolor' (bicolor monkshood) *Aconitum carmichaelii* 'Pink Sensation,' 125
Aconitum episcopale, 210
Aconitum lamarckii, 124, 125
Aconitum napellus, 127
Aconitum x *cammarum* 'Bicolor' (bicolor monkshood), *124,* 124–25
Actaea simplex 'Hillside Black Beauty' (black bugbane), *94, 121, 140*
Actinidia kolomikta 'Arctic Beauty' (arctic kiwi; hardy kiwi vine), 210–11
African daisy (*Osteospermum*), *104,* 104–5
Ajuga, 172
Ajuga reptans 'Burgundy Glow' (bugleweed), *6, 24*
Alaska State Fair Vine (*Rhodochiton atrosanguineum*), 105
Alchemilla mollis (lady's mantle), *57, 117, 151*
Allium aflatunense 'Purple Sensation' (ornamental onion), 108, *108 Allium cernuum* (nodding onion), 84, 125–26, *126*
Allium sphaerocephalon (drumstick allium), *39, 43*
Alopecurus pratensis 'Aureovariegatus' (golden foxtail grass; golden meadow foxtail), 116, *116*
Alopecurus pratensis 'Variegatus' (golden foxtail grass; golden meadow foxtail), 116
See also *Alopecurus pratensis* 'Aureovariegatus' (golden foxtail grass; golden meadow foxtail)
alpine bells, *156*
alpine clematis (*Clematis alpina* 'Constance'), 211–12
alyssum, bees and, 14
amaranth, *101*

Amaranthus caudatus 'Love-lies-bleeding,' 102, *102*
Amaranthus hypochondriacus 'Chinese Giant Orange' ('Chinese Giant Orange' amaranth), 102
American Horticultural Society
heat zone map, 66
American mountain ash (*Sorbus Americana*), *22*
Amur chokecherry (*Prunus maackii*), 17, 204–5
Amur maple (*Acer ginnala*; *Acer tataricum subsp. ginnala*), 195–96
Anemone narcissiflora (narcissus-flowered anemone), *37*
annuals, 101–5
shopping tips, 83
See also names of specific annuals
Anticlea elegans (camas wand lily; death camas), 177
See also *Zigadenus elegans* (camas wand lily; death camas)
aphids, 80
architectural shape and form, plant, 25–26
conifers, 25
ornamental grasses, 25
arctic kiwi (*Actinidia kolomikta* 'Arctic Beauty'), 210–11
See also hardy kiwi vine (*Actinidia kolomikta* 'Arctic Beauty')
Armenian grape hyacinth (*Muscari armeniacum*), 110
Artemisia ludoviciana 'Valerie Finnis' (western mugwort; white sage), 126–27, *127, 171*
artichoke thistle (*Cynara cardunculus*), architecture of, 26
Asclepias spp (milkweed), *47*
ash leaf spirea Sem (*Sorbaria sorbifolia* 'Sem'), 188–89
Asiatic lilies, 14, *136, 151*
Asiatic primrose (*Primula alpicola*; *Primula waltonii*), 159–60
Astilboides tabularis (shieldleaf), 128, *128*
See also *Rodgersia astilboides* (shieldleaf)
Astrantia major 'Alba,' 129
Astrantia major 'Claret,' 129
Astrantia major 'Florence,' 129
Astrantia major 'Hadspen Blood' (Hattie's Pincushion; masterwort), *43,* 128–29, *129, 134, 174*
Astrantia major 'Lars,' 129
Astrantia major 'Moulin Rouge,' 129
Astrantia major 'Pink Pride,' 129
Astrantia major 'Ruby Wedding,' 129
Astrantia major 'Shaggy,' 129
Astrantia major 'Star of Beauty,' *128*
Astrantia major 'Star of Billion,' *129*
Astrantia major 'Tickled Pink,' 129
Astrantia major 'White Giant,' 129
auricula primrose (*Primula auricular*), 157–58, *158*
autumn fire stonecrop (*Sedum spectabile* 'Autumn Fire'), 169–70
avens (*Geum* 'Tim's Tangerine'; *Geum* 'Totally Tangerine'), *42,* 138
axillary shoot, 147, 215

bachelor's button (*Centaurea montana*), *1*
Bailey compact cranberry (*Viburnum trilobum* 'Bailey Compact'), 192–93
Bailey Select Schubert Cherry (*Prunus virginiana* 'Bailey Select'), 205–07
bare root, 93, 215
  plants, *94*
bark, 17
  fall, 37
  winter, 41
  See also names of specific trees
bear's ear (*Primula auricular*), 157–58
  See also auricula primrose (*Primula auricular*)
bee balm, bees and, 14, 46
bees, 45–46
  bee balm and, 14, 46
  flowers and, 14
  herbaceous perennials and, 132, 150, 169, 173
  See also pollinators
begonias, 107
behavior, plant, 35, 51–53, 57
Berberis thunbergii (Japanese barberry), 179–82
  See also names of plants in the Berberis thunbergii/barberry family
Berberis thunbergii 'Aurea' (golden barberry), 179–80
Berberis thunbergii 'Orange Rocket' (orange rocket barberry), *181,* 181–82
Berberis thunbergii var. *atropurpurea,* 98
Berberis thunbergii var. *atropurpurea* 'Crimson Pygmy' ('Crimson Pygmy' barberry), 180
Berberis thunbergii var. *atropurpurea* 'Crimson Velvet'(crimson velvet barberry), *180,* 180–81, *181*
Berberis thunbergii var. *atropurpurea* 'Rose Glow' (rose glow barberry), 182, *182*
Bergenia 'Bressingham Ruby,' *127*
berries, 21, *22,* 215
  fall, 37
  shrub, 180, 182, 183, 192
  tree, 195, 199, *201,* 206
  winter, 41
  See also names of berry-bearing plants
Betula nigra (river birch), 89
Betula papyrifera (canoe birch; paper birch; white birch), 197–98, *198*
bicolor monkshood (*Aconitum* x *cammarum* 'Bicolor'), *124,* 124–25
bidens (*Bidens ferulifolia*), 103, *103*
Bidens ferulifolia (bidens; tickseed), 103
bigleaf Ligularia (*Ligularia dentate*), 150–51
birds
  plant motion and, 29
  seed-eating, 150
  shrubs and, 184
  tree berries and, 206
  See also hummingbirds
bitterroot (*Lewisia*), 148–49

black bugbane (*Actaea simplex* 'Hillside Black Beauty'), *94, 121, 140*
black-eyed Susans (*Rudbeckia fulgida*), 66
black iris (*Iris chrysographes* 'Black Form'), 143
Blaze of Fulda stonecrop (*Sedum spurium* 'Blaze of Fulda'), 168–69
blazing star (*Liatris spicata* 'Kobold'), 149–50
bloody cranesbill (*Geranium sanguineum* 'Vision Violet'), 136–37
bloom time and length, 35, 43
blue honeywort (*Cerinthe major* 'Purpurascens'), *103,* 103–4
blue kiwi (*Cerinthe major* 'Purpurascens'), 103–4
  See also blue honeywort (*Cerinthe major* 'Purpurascens')
blue oat grass (*Helictotrichon sempervirens*), 121, *121*
blue spruce (*Picea pungens* 'Fat Albert'), *24*
botanical (scientific) names, 97–98
  See also cultivar; genus; hybrid symbol; species; subspecies; variety
Bracteantha bracteata 'Mohave Deep Rose' (straw flowers), *101*
bridal wreath spirea, 191
broad-leaved grape hyacinth (*Muscari latifolium*), 110
  See also *Muscari latifolium* (broad-leaved grape hyacinth)
bronze veil tufted hair grass (*Deschampsia cespitosa* 'Bronzeschleier'), 119–20
bronzeleaf Rodgersia (*Rodgersia podophylla* 'Rotlaub'), 163–64
bugbane. See *Actaea simplex* 'Hillside Black Beauty'
bugleweed (*Ajuga reptans* 'Burgundy Glow'), *6, 24*
built environments, microclimates and, 63
bulbs, 107–13
  corms, 107
  rhizomes, 107
  shopping tips, 83–84
  spring, 37
  susceptibility to burrowing animals, 83–84
  true bulbs, 107
  tubers, 107
  tunic, 83
  where to plant, 83
  See also names of specific bulbs
Burbank, Luther, 147
butter and eggs (*Linaria vulgaris*), 51
butterflies, 45
  flowers and, 14
  herbaceous perennials and, 150, 169, 173, 176
  shrubs and, 182, 189
buying plants
  in bud versus in bloom, 79–80
  late bloomers, 93–95
  local versus online, 77
  researching before, 95
  root structure and, 93, 95
  selecting best for your garden, 79–81
  See also plant shopping tips

Calamagrostis x *acutiflora* (feather reed grass), 116
Calamagrostis x *acutiflora* 'Avalanche' (variegated feather reed grass), 118–19, *119*
Calamagrostis x *acutiflora* 'Eldorado' PP16,486 (golden feather reed grass), 117, *117, 118*
Calamagrostis x *acutiflora* 'Karl Foerster' (feather reed grass), *41,* 117–18, *118*
Calamagrostis x *acutiflora* 'Overdam' (variegated feather reed grass), *29,* 118–19, *119*
calendula, *100*
calyx, 216
camas wand lily (*Zigadenus elegans*), 177
Camassia quamash (wild hyacinth), 177
Campanula persicifolia 'La Belle,' 14, 86, *86*
Canadian Department of Natural Resources
  hardiness zone map, 66
canoe birch (*Betula papyrifera*), 197–98
  See also paper birch (*Betula papyrifera*); white birch (*Betula papyrifera*)
Cape daisy (*Osteospermum*), 105
'Caradonna' salvia (*Salvia nemorosa* 'Caradonna'), *128*
cardoon, architecture of, 26
carex, 115
caterpillars, Eastern tent
  trees and, 206
catmint (*Nepeta* x *faassenii* 'Six Hills Giant'; *Nepeta* x *faassenii* 'Walker's Low'), 42, *42, 128,* 153–54
  bees and, 14, 46
cattails, 115
Centaurea montana (bachelor's button), *1*
Centaurea montana (giant blue bachelor's buttons), 51
central leader, 89, *89,* 215
Cephalaria tatarica (giant yellow scabious), *44*
Cerastium tomentosum (snow-in-summer), 14
Cerinthe major 'Purpurascens' (blue honeywort) 103–4
Cerinthe major var. purpurascens (blue honeywort), 103–4
chard ('Bright Lights'), *19*
checkered lily (*Fritillaria meleagris*), 109
Chinese giant orange amaranth (*Amaranthus hypochondriacus* 'Chinese Giant Orange'), 102, *102*
Chionodoxa forbesii (glory-of-the-snow), 108–9
  See also *Chionodoxa forbesii* 'Blue Giant' (glory-of-the-snow)
Chionodoxa forbesii 'Blue Giant' (glory-of-the-snow), 108, *108*
Chionodoxa forbesii 'Pink Giant' (glory-of-the-snow), 108
Chionodoxa luciliae (glory-of-the-snow), 108–9
Chionodoxa siehei (glory-of-the-snow), 108–9
Clematis (alpine clematis; clematis; old man's beard; Russian virgin's bower), 211–13
  See also names of plants in the Clematis family
clematis (*Clematis*), 211–13
Clematis alpina 'Constance' (alpine clematis), 211–12, *212*
Clematis alpine 'Willy," 212

Clematis integrifolia (shrub clematis; solitary clematis), 130, *130*
Clematis tangutica (Russian virgin's bower), *90,* 212–13, *213*
cliff maids (*Lewisia cotyledon*), 148
cliff stonecrop (*Sedum cauticola*), 169
climbing monkshood (*Aconitum aff. hemsleyanum*; *Aconitum aff. hemsleyanum* 'Red Wine'), *208, 209,* 209–10
color
  flowers, 13
  foliage, 9
columbine (*Aquilegia* 'Colorado Violet and White'), 14, *43*
  bees and, 46
compact American highbush cranberry (*Viburnum trilobum* 'Bailey Compact'), 192–93
compost, 67
conifers, *89*
  architectural shape and form, 25
  cones, 21, *23*
  evergreen, 41
coral carpet stonecrop (*Sedum album* 'Coral Carpet'), 168
Coreopsis verticillata 'Zagreb' (threadleaf coreopsis), *34*
corms, 107
Cornus alba (red twig dogwood), 182–84
  See also names of plants in the Cornus alba/dogwood family
Cornus alba 'Bailhalo' (variegated red twig dogwood), 183, *183*
Cornus alba 'Elegantissima' (variegated red twig dogwood), 183, *183*
Cornus alba 'Gouchaultii' (Tatarian dogwood), *183,* 183–84
Cornus sericea 'Flaviramea' (yellow-twigged dogwood), 41
corymb, 215
  shrub, 186, 192
cowslip (*Primula*), 156–61
  See also primrose (*Primula*)
cranesbill (*Geranium*), 136–37
Crataegus x *mordenensis* 'Toba' (Toba hawthorn), 199–200
creeping thyme (*Thymus praecox* 'Highland Cream'; *Thymus serpyllum* 'Pink Chintz'), 172, 173
'Crimson Pygmy' barberry (*Berberis thunbergii* var. *atropurpurea* 'Crimson Pygmy'), 180, *180*
crimson velvet barberry (*Berberis thunbergii* var. *atropurpurea* 'Crimson Velvet'), 180–81
Crocus vernus 'Flower Record,' *38*
crocuses, *38,* 107
crown, 80, 215
  perennials, 148
  shrub, 88
  tree, 196, 198
cultivar, 97, 98, 215
  See also names of specific plants
cultural requirements of plants, 65–69
  day and season lengths, 65

humidity, 65
moisture, 65, 67, 69
soil fertility, 65, 67
soil pH, 65, 67
soil type, 65, 67
sunlight, 65
temperature range, 65, 66–67
wind exposure, 65
See also hardiness zones; heat zones; soil
Culver's root (*Veronicastrum virginicum* 'Apollo'), 176
motion, 29
cushion spurge (*Euphorbia polychrome*; *Euphorbia polychroma* 'Bonfire'), *11,* 19, *25*
cyme, 215
shrub, 183

daffodil (*Narcissus*), 14, 84, *85,* 107, *111,* 111–12, *116*
daisies, 14, *52,* 86
butterflies and, 14
dame's rocket (*Hesperis matronalis*), *49*
dappled shade, 65
day lilies, 86
deadheading, 37, 57, 137
annuals, 83, 104
death camas (*Zigadenus elegans*), 177
delphinium, *56*
bees and, 46
Deschampsia cespitosa 'Bronzeschleier' (bronze veil tufted hair grass), 119–20, *120*
Dianthus 'Coconut Surprise,' 33
Dianthus gratianopolitanus 'Firewitch' (garden pink), *4*
dill, *20*
disease susceptibility, plant, 57
dividing, plant, 57
Dodecatheon dentatum (white shooting star), 131, *131*
Dodecatheon pulchellum ssp. *alaskanum* (shooting stars), 131, *131*
Doronicum 'Little Leo' (leopard's bane), *39*
drought-tolerant plants, 67, 69
See also names of specific types of plants
drumstick allium (*Allium sphaerocephalon*), *39, 43*
drumstick primroses, *156*
drupe, 215
shrub, 181, 184
tree, 195
dwarf Alberta spruce (*Picea glauca* 'Conica'), 202–3
dwarf arctic iris (*Iris setosa*), 145
dwarf arctic willow (*Salix purpurea* 'Nana'), 17, 19, *187,* 187–88
dwarf daffodil, *111*
dwarf globeflower (*Trollius pumilus*), 174
dwarf meadowsweet (*Filipendula* 'Kahome'), *24*

dwarf purple osier (*Salix purpurea* 'Nana'), 187–88
  See also dwarf arctic willow (*Salix purpurea* 'Nana')
dwarf trollius (*Trollius pumilus*), 174

early-season display plants, 35, 37
  deadheading, 37
Echinacea purpurea (purple coneflowers), 66
Echinops ritro 'Veitch's Blue' (globe thistle), 132, *132*
environment, understanding growing, 62–63
  See also built environments, microclimates and; cultural requirements of plants; hardiness zones; heat zones; light and lighting; soil; wind conditions
Eryngium alpinum, 133
Eryngium bourgatii, 133
Eryngium 'Sapphire Blue,' 133, *133*
Eryngium x *zabelii* 'Big Blue' (sea holly), 132–33, *133*
Euphorbia polychroma (cushion spurge), 19, *25*
Euphorbia polychroma 'Bonfire' (cushion spurge), *11*
European meadowsweet (*Filipendula ulmaria* 'Variegata'), 135–36
evergreen conifers, 41
fall-display plants, 35, 37
false spirea Sem (*Sorbaria sorbifolia* 'Sem'), 188–89
farinose, 215
  perennials, 159, 160
'Fat Albert' Colorado spruce (*Picea pungens* 'Fat Albert'), *79*
feather reed grass (*Calamagrostis* x *acutiflora*; *Calamagrostis* x *acutiflora* 'Karl Foerster'), *115,* 116, 117–18
  architecture, 26
fiddlehead ferns, 37
Filipendula (meadowsweet), 134–36
  See also plants in the Filipendula/meadowsweet family
Filipendula 'Kahome' (dwarf meadowsweet), *24*
Filipendula purpurea 'Elegans' (Japanese meadowsweet), 134, *134*
Filipendula rubra 'Venusta' (queen of the prairie), *124,* 134–35, *135*
Filipendula ulmaria, nonvariegated, *135*
Filipendula ulmaria 'Variegata' (European meadowsweet), *135,* 135–36
fingerleaf Rodgersia (*Rodgersia henrici* 'Cherry Blush'), 164–65
First Editions Superstar Spirea (*Spiraea* x *bumalda* 'Denistar'), 189–90
flowering crab apple (*Malus* 'Royalty'), 201–2
flowers, 13–14
foliage, 9–10
forget-me-nots (*Myosotis*), *37, 117*
foxgloves, bees and, 14
foxtail barley (*Hordeum jubatum*), 116
fragrance, plant, 31–33
Fritillaria meleagris (checkered lily; guinea hen flower; snake's head), 84, 109, *109*
Fritillaria meleagris 'Aphrodite,' 109, *109*
Fritillaria meleagris 'Charon,' 109
Fritz Creek Gardens, 2

full shade, 65
full sun, 65

garden pink (*Dianthus gratianopolitanus* 'Firewitch'), *4*
garden sage (*Salvia nemorosa*), 165
gayfeather (*Liatris spicata* 'Kobold'), 149–50
gentian, *92*
Gentiana lutea (great yellow gentian), *25*
genus, 97, 98, 215
  See also names of specific plants
Geranium (cranesbill; hardy geranium), 136–37
  See also names of plants in the Geranium/cranesbill family
geranium (*Geranium* 'Sabani Blue'), *49*
Geranium sanguineum 'Vision Violet' (bloody cranesbill; hardy geranium), *136,* 136–37
Geranium wallichianum 'Rozanne,' 136
Geranium x *magnificum* (showy geranium), 137, *137*
Geum 'Tim's Tangerine' (avens), 138
  See also *Geum* 'Totally Tangerine' (avens)
Geum 'Totally Tangerine' (avens), *42,* 138, *138*
giant blue bachelor's buttons (*Centaurea montana*), 51
giant yellow scabious (*Cephalaria tatarica*), *44*
globe thistle (*Echinops ritro* 'Veitch's Blue'), 14, 132
globeflower (*Trollius*), 173–75
  See also *Trollius chinensis* 'Golden Queen' (globeflower; queen of the buttercups)
glory-of-the-snow, 84, *108,* 108–09
golden barberry (*Berberis thunbergii* 'Aurea'), *178,* 179–80
golden feather reed grass (*Calamagrostis* x *acutiflora* 'Eldorado' PP16,486), 117
golden foxtail grass (*Alopecurus pratensis* 'Aureovariegatus'), *42,* 116
golden meadow foxtail (*Alopecurus pratensis* 'Aureovariegatus'), 116
golden ninebark (*Physocarpus opulifolius* 'Dart's Gold'), 185–86
grape hyacinth (*Muscari*), 109–11
  See also names of plants in the Muscari/hyacinth family
grasses, 115–21
  carex, 115
  cattails, 115
  clumping, 84
  cool-season, 41
  cold season, 84, 86
  inflorescences, 115
  running, 84
  rushes, 115
  sedges, 115
  shopping tips, 84, 86
  warm season, 84, 86
  See also names of specific grasses
great yellow gentian (*Gentiana lutea*), *25*

green lacewings, 80
guinea hen flower (*Fritillaria meleagris*), 109

Hakonechloa macra 'Aureola' (Japanese fountain grass), 116
half-hardy perennial, 215
See also blue honeywort; blue kiwi; *Cerinthe major* 'Purpurascens'; *Cerinthe major var. purpurascens)*
Hamamelis (witch hazel), 188
hardening off, 72
hardiness, plant, 66
hardiness zones, 63, 66–67
experiments in pushing, 66
hardy geranium (*Geranium*), 136–37
See also *Geranium sanguineum* 'Vision Violet' (bloody cranesbill; hardy geranium)
hardy kiwi vine (*Actinidia kolomikta* 'Arctic Beauty'), *210,* 210–11
Hattie's Pincushion (*Astrantia major* 'Hadspen Blood'), 128–29
See also masterwort (*Astrantia major* 'Hadspen Blood')
heat zones, 63, 66
Helictotrichon sempervirens (blue oat grass), 121, *121*
herbaceous perennials, 19
early bloomers, 87
late bloomers, 87
life expectancy, 123
mid-season bloomers, 87
shopping tips, 86–87
size expectations, 49
stems, 19
See also names of specific perennials; perennials
Hesperis matronalis (dame's rocket), *49*
Himalayan blue poppy (*Meconopsis betonicifolia*), 53, 66, 152–53
Hippophae rhamnoides (sea berry; seabuckthorne), 200–1
hosta (*Hosta*), 139–42, *151*
Hosta (hosta; plantain lily), 139–42
See also names of plants in the Hosta family
Hosta 'Blue Mouse Ears,' 140, *140,* 141
Hosta 'Canadian Blue,' *140,* 140–41
Hosta 'Gold Standard,' 141, *141*
Hosta 'Krossa Regal,' *139*
Hosta 'Patriot,' *10,* 141, *141*
Hosta 'Regal Splendor,' *139,* 142, *142*
hummingbirds, 45, 46, 150
hybrid symbol, 99

Iberis sempervirens 'Alexander's White,' *86*
inflorescences, 115, *115, 120,* 215
insects
attraction to plants, 57

beneficial, 80–81
damaging, 80
See also names of specific types of insects; bees; butterflies; pollinators
integrated pest management (IPM), 80
Iris (iris), 142–46
See also names of plants in the Iris family
iris (*Iris*), 86, 142–46
Iris arenaria (sand iris), 143, *143*
Iris chrysographes 'Black Form' (black iris), 143, *143*
Iris pseudacorus (yellow flag iris), 142
Iris pseudacorus 'Sun Cascade' (sun cascade yellow flag iris), 143–44, *144*
Iris pseudacorus 'Variegata,' *144*
Iris pumila (standard dwarf bearded iris), 142, 144–45
Iris pumila 'Betsy Boo,' 145
Iris pumila 'Candy Apple,' *144,* 145
Iris pumila 'Mauhaus,' 145
Iris pumila 'Navy Doll,' 145
Iris setosa (dwarf arctic iris; native iris; wild iris), *117,* 142, 145, *145*
Iris sibirica (Siberian iris), 142, 145–46
Iris sibirica 'Caesar's Brother,' 146, *146*
Iris sibirica 'Ruffled Velvet,' 146, *146*
Iris sibirica 'Silver Edge,' 146, *146*

Japanese barberry (*Berberis thunbergii*), 179–82
Japanese coltsfoot (*Petasites japonicus* var. giganteus 'Variegata'), *55*
Japanese fountain grass (*Hakonechloa macra* 'Aureola'), 116
Japanese meadowsweet (*Filipendula purpurea* 'Elegans'), 134
Juliana hybrids (*Primula* x *juliae*), 160–61

lady's mantle (*Alchemilla mollis*), *57, 117, 151*
lambs' ears (*Stachys byzantina* 'Helen von Stein'), *6,* 9
bees and, 14
Lamium maculatum 'White Nancy' (spotted deadnettle), *10*
lavender, *60, 61*
bees and, 46
leader, 89
leopard's bane (*Doronicum* 'Little Leo'), *39*
Leucanthemum vulgare (oxeye daisy), 51, 147
Leucanthemum x *superbum* 'Alaska,' 147
Leucanthemum x *superbum* 'Banana Cream' (Shasta daisy), *121,* 147, *147*
Lewisia (bitterroot; Lewisia), 148–49
See also names of plants in the Lewisia family
Lewisia (*Lewisia*), 148–49
Lewisia cotyledon (cliff maids), 148
Lewisia cotyledon 'Little Peach,' *148*
Lewisia cotyledon 'Little Plum,' *148*
Lewisia tweedyi (Tweedy's Lewisia), *68,* 148–49, *149*
Lewisiopsis tweedyi (Tweedy's Lewisia), 148–49

See also *Lewisia tweedyi* (Tweedy's Lewisia)
Liatris spicata 'Floristan Violet,' 150
Liatris spicata 'Kobold' (blazing star; gayfeather), 149–50, *150*
light and lighting, 1, 3
amount, 62
Ligularia, 150–52
See also names of plants in the Ligularia family
Ligularia dentate (bigleaf Ligularia), 150–51
Ligularia dentata 'Britt-Marie Crawford,' 151
Ligularia dentata 'Desdemona,' 151
Ligularia dentata 'Othello,' 151
Ligularia dentata x *hessei* 'Gregynog Gold,' 151, *151*
Ligularia denticulata 'Othello,' *34*
Ligularia przewalskii, 151
Ligularia stenocephala (narrow-spiked Ligularia), 150, 151–52
Ligularia stenocephala 'Bottle Rocket,' 152
Ligularia stenocephala 'Little Rocket,' 152
Ligularia stenocephala 'The Rocket,' *48, 151,* 151–52
lilac (*Syringa vulgaris* 'Sensation'), *30*
Lilium LA Hybrid 'Royal Sunset,' *13*
Linaria vulgaris (butter and eggs), 51
Linnaeus, Carl, 97
London pride saxifrage (*Saxifraga* x urbium 'Aurea punctata'), *41*
lungwort (*Pulmonaria* 'Trevi Fountain'), *10*

maintenance, plant, 35
ease of, 57
See also behavior, plant
Malus 'Dolgo,' 201
Malus 'Makamik,' 201–2
Malus 'Royalty' (flowering crab apple), *201,* 201–2
Malus 'Thundercloud,' 202
mango tango cinquefoil (*Potentilla fruticosa* 'Mango Tango'), 186–87
See also *Potentilla fruticosa* 'UMan' (mango tango cinquefoil)
marginals, 69
masterwort (*Astrantia major* 'Hadspen Blood'), 43, 128–29
meadowsweet (*Filipendula*), 134–36
See also names of plants in the meadowsweet/Filipendula family
Meconopsis betonicifolia (Himalayan blue poppy), *152,* 152–53
milkweed (*Asclepias* spp.), *47*
'Miss Kim' lilac (*Syringa pubescens* subsp. *patula* 'Miss Kim'), *40*
mites, predatory, 80
mockorange (*Philadelphus lewisii* 'Blizzard'), 184–85
monkshood (*Aconitum napellus*), common, 14, 86, 125
bees and, 46
self-sowing, 125
moss saxifrage (*Saxifraga* x *arendsii*), 166–67
mossies (*Saxifraga* x *arendsii*), 166–67

motion, plant, 29
mountain cowslip (*Primula auricular*), 157–58
See also Auricula primrose (*Primula auricular*)
mullein (*Verbascum*), *18*
Muscari (grape hyacinth), 109–11
See also names of plants in the Muscari/grape hyacinth family
Muscari armeniacum (Armenian grape hyacinth), *109,* 110
Muscari latifolium (broad-leaved grape hyacinth), 110, *110*
Muscari neglectum 'Valerie Finnis' (Valerie Finnis grape hyacinth), 110–11

Narcissus (daffodil), 111–12
Narcissus 'Manly,' *33, 107, 111*
narcissus-flowered anemone (*Anemone narcissiflora*), *37*
narrow-spiked Ligularia (*Ligularia stenocephala*), 151–52
native iris (*Iris setosa*), 145
naturalizes, 215
See also names of specific plants
nematodes, beneficial, 80
Nepeta cataria, 154
Nepeta faassenii 'Six Hills Giant,' *154*
Nepeta faassenii 'Walker's Low,' *117, 154*
Nepeta racemosa 'Walker's Low' (catmint), 153–54
See also *Nepeta* x *faassenii* 'Walker's Low' (catmint)
Nepeta 'Six Hills Giant' (catmint), 153–54
See also *Nepeta* x *faassenii* 'Six Hills Giant' (catmint)
Nepeta x *faassenii* 'Six Hills Giant' (catmint), 43, 153–54
See also *Nepeta faassenii* 'Six Hills Giant'
Nepeta x *faassenii* 'Walker's Low' (catmint), 43, 153–54
See also *Nepeta faassenii* 'Walker's Low'
new moon globeflower (*Trollius* x *cultorum* 'New Moon'), 175
nine bark (*Physocarpus opulifolius* 'Diablo'), *88*
nodding onion (*Allium cernuum*), *96, 125,* 125–26
Northern Jacob's ladder (*Polemonium boreale* 'San Juan Skies'), *47,* 155
nurseries, 71–72
criteria for choosing, 75
integrated pest management (IPM) and, 80
See also buying plants; plant shopping tips

old man's beard (*Clematis*), 211–13
Olea europaea, 200
orange rocket barberry (*Berberis thunbergii* 'Orange Rocket'), 181–82
ornamental grasses, architectural shape and form of, 25
ornamental onion (*Allium aflatunense* 'Purple Sensation'), 14, 108
bees and, 14
ornamental rhubarb (*Rheum palmatum* 'Atrosanguineum'), 162–63
Osteospermum (African daisy; Cape daisy), *104,* 104–5
oxeye daisy (*Leucanthemum vulgare*), 51, 147

painted daisy (*Pyrethrum coccineum*; *Tanacetum coccineum* 'Robinson's Hybrid'), *49*, 171, *171*
panicle, 136, 215
See also names of specific plants
pansies, 83
paper birch (*Betula papyrifera*), 197–98, *198*
See also white birch (*Betula papyrifera*)
part shade, 65
part sun, 65
pasque flower (*Pulsatilla vulgaris* 'Red Bells'; *Pulsatilla vulgaris* 'Violet Bells'), 21, 161–62
"pass-around plants," 51
patented plant, 215–16
See also *Calamagrostis* x *acutiflora* 'Eldorado' PP16,486 (golden feather reed grass); *Eryngium* x *zabelii* 'Big Blue' (sea holly); golden feather reed grass (*Calamagrostis* x *acutiflora* 'Eldorado' PP16,486)
peonies, *94, 123*
bees and, 46, *46*
perennials, 86, 123
well-known examples, 86
See also names of specific perennials; herbaceous perennials
Pericallis 'Senetti' ('Senetti'), *101*
Petasites japonicus var. giganteus 'Variegata' (Japanese coltsfoot), *55*
Phalaris arundinacea 'Picta' (ribbon grass), *50*
Philadelphus lewisii 'Blizzard' (mockorange), *33, 184,* 184–85
Phlox subulata 'Candy Stripes,' *14*
Phlox subulata 'Emerald Pink,' *14*
Physocarpus opulifolius 'Dart's Gold' (golden ninebark), *185,* 185–86
Physocarpus opulifolius 'Diablo' (nine bark), *88*
Picea glauca 'Conica' (dwarf Alberta spruce), *202,* 202–3
Picea glauca 'Sander's Blue,' 203
Picea pungens 'Fat Albert' (blue spruce), *24*
pink parasols spirea (*Spiraea fritschiana* 'Pink Parasols'), 190–91
See also *Spiraea fritschiana* 'Wilma' (pink parasols spirea)
plant qualities, desirable artistic, 2
See also architectural shape and form, plant; bark; berries; flowers; foliage; fragrance, plant; motion, plant; seedheads; seedpods; stems
plant qualities, desirable non-artistic, 35
See also behavior, plant; bloom time and length; early-season display plants; fall-display plants; maintenance, plant; pollinators; shoulder-season plants; size and scale, plant; suitability, plant; vigor, plant; winter-display plants
plant shopping tips, 83–91
annuals, 83
bulbs, 83–84
grasses, 84, 86
herbaceous perennials, 86–87
shrubs, 87–88
trees, 88–89, 91
vines, 91

plantain lily (*Hosta*), 139–42
See also hosta (*Hosta*)
plugs, 93, 216
poker primrose, *157*
Polemonium boreale 'San Juan Skies' (northern Jacob's ladder), 155, *155*
pollinators, 2, 14, 216
attraction to plants, 37, 45–47
creating habitats for, 46
plant motion and, 29
See also bees; butterflies; hummingbirds
poplar (*Populus tremuloides*), 203–4
See also quaking aspen (*Populus tremuloides*)
poppy, 86, *153*
Populus tremuloides (poplar; quaking aspen), *203,* 203–4
Potentilla fruticosa 'Katherine Dykes,' 187, *187*
Potentilla fruticosa 'Mango Tango' (mango tango cinquefoil), *186,* 186–87
Potentilla fruticosa 'Pink Beauty,' *186,* 187
Potentilla fruticosa 'UMan' (mango tango cinquefoil), 186–87
See also *Potentilla fruticosa* 'Mango Tango' (mango tango cinquefoil)
praying mantis, 80
primrose (*Primula*), *125, 131,* 156–61
resources, 156
Primula (cowslip; primrose), 156–61
resources, 156
See also names of plants in the Primula/cowslip/primrose family
Primula alpicola (Asiatic primrose), 159–60, *160*
Primula auricular (auricula primrose; bear's ear; mountain cowslip), 157–58, *158*
Primula cockburniana, 157
Primula denticulate 'Ronsdorf Strain,' *156*
Primula florindae (Tibetan primrose), *41, 119,* 127, *158,* 158–59
Primula marginata (silver-edged primrose), 159, *159*
Primula matthioli, 156
Primula vialii, 157
Primula waltonii (Asiatic primrose), 159–60, *160*
Primula x *juliae* (Juliana hybrids), 160–61
Primula x *juliae* 'Dorothy,' 161, *161*
Primula x *juliae* 'Wanda,' 160–61, *161*
Primula x *pruhoniciana* (Juliana hybrids), 160–61
pruning, 57
Prunus maackii (Amur chokecherry), *204,* 204–5, *205*
Prunus virginiana 'Bailey Select' (Bailey Select Schubert Cherry), 205–7, *206, 207*
Pulmonaria 'Trevi Fountain' (lungwort), *10*
Pulsatilla vulgaris 'Red Bells' (Pasque flower), 161–62, *162*
See also *Pulsatilla vulgaris* 'Violet Bells' (Pasque flower)
Pulsatilla vulgaris 'Violet Bells' (Pasque flower), *161,* 161–62
See also *Pulsatilla vulgaris* 'Red Bells' (Pasque flower)
purple bell vine (*Rhodochiton atrosanguineum*), 105
purple coneflowers (*Echinacea purpurea*), 66
purple emperor stonecrop (*Sedum spectabile* 'Purple Emperor'), 170

Pyrethrum coccineum (painted daisy), 171, *171*

quaking aspen (*Populus tremuloides*), *16, 29, 203,* 203–04
queen of the buttercups (*Trollius chinensis* 'Golden Queen'), 174
See also globeflower
queen of the prairie (*Filipendula rubra* 'Venusta'), 134–35
motion, 29

raceme, 111, 175, 176, 216
red creeping thyme (*Thymus serpyllum* 'Coccineus'), 173
red Norway maple (*Acer platanoides* 'Royal Red'), 196–97
red rose hips (*Rosa acicularis*), *41*
red twig dogwood (*Cornus alba*), *17,* 182–84
redleaf Rodgersia (*Rodgersia podophylla* 'Rotlaub'), 163–64
red-twigged dogwood (*Cornus sericea* 'Cardinal'), 41
Rheum palmatum 'Atrosanguineum' (ornamental rhubarb), *162,* 162–63
architecture, 26
rhizome, 51, 107, 127, 145, 216
See also names of specific plants
Rhodochiton atrosanguineum (Alaska State Fair vine; purple bell vine), 105, *105*
ribbon grass (*Phalaris arundinacea* 'Picta'), *50*
river birch (*Betula nigra*), 89
rock soapwort (*Saponaria ocymoides*), *119*
rocket ligularia (*Ligularia stenocephala* 'The Rocket'), 19
Rodgersia (Roger's flower), 163–65
See also names of plants in the Rodgersia/Roger's flower family
Rodgersia astilboides (shieldleaf), 128, 165
Rodgersia henrici 'Cherry Blush' (fingerleaf Rodgersia), *164,* 164–65
Rodgersia podophylla 'Rotlaub' (bronzeleaf Rodgersia; redleaf Rodgersia), *11, 163,* 163–64, *164*
Roger's flower (*Rodgersia*), *11,* 163–65
root-bound, 79, 216
Rosa acicularis (red rose hips), *41,* 53, *53*
rose glow barberry (*Berberis thunbergii* var. *atropurpurea* 'Rose Glow'), 182
royal red maple (*Acer platanoides* 'Royal Red'), 196–97
rushes, 115
Russian virgin's bower (*Clematis*), 211–13
See also Clematis tangutica

sage, bees and, 14, 46
Salix purpurea 'Gracilis' (dwarf arctic willow; dwarf purple osier), 187–88
Salix purpurea 'Nana' (dwarf arctic willow; dwarf purple osier), 187–88
salvia, 14
Salvia nemorosa (garden sage), 165
Salvia nemorosa 'Caradonna' ('Caradonna' salvia), *128*
Salvia nemorosa 'May Night' (wood sage), 165, *165*
sand iris (*Iris arenaria*), 143, *143*
Saponaria ocymoides (rock soapwort), *119*

Saxifraga apiculata 'Gregor Mendel,' *166*, 167
Saxifraga x *arendsii* (moss saxifrage; mossies), 166–67
Saxifraga x urbium 'Aurea punctata' (London pride saxifrage), *41*
scape, 216
See also names of specific plants
scarlet-stemmed meadowsweet (*Filipendula purpurea* 'Elegans'), 19
Scilla (spring squill), 112–13
See also names of plants in the Scilla/squill family
Scilla mischtschenkoana (Tubergeniana squill), 112, *112*
Scilla siberica (Siberian squill), 113, *113*
Scilla tubergeniana (Tubergeniana squill), 112
See also *Scilla mischtschenkoana* (Tubergeniana squill)
sea berry (*Hippophae rhamnoides*), 200–1
See also seabuckthorne (*Hippophae rhamnoides*)
sea holly (*Eryngium* x *zabelii* 'Big Blue'), *125*, 132–33, *133*
seabuckthorne (*Hippophae rhamnoides*), *200*, 200–1
sedges, 115
sedum (*Sedum*), *38*, 167–70
Sedum (sedum; stonecrop), *68*, *125*, *167*, 167–70, *168*
See also names of plants in the Sedum/stonecrop family
Sedum album 'Coral Carpet' (coral carpet stonecrop), 168, *168*
Sedum cauticola (cliff stonecrop), 169, *169*
Sedum kamtschaticum (stonecrop), *47*, 167
Sedum spectabile 'Autumn Fire' (autumn fire stonecrop), 169–70, *170*
Sedum spectabile 'Purple Emperor' (purple emperor stonecrop), 170, *170*
sedum 'Spice of Life,' *27*
Sedum spurium 'Blaze of Fulda' (blaze of Fulda stonecrop), 168–69, *169*
Sedum telephium 'Purple Emperor' (purple emperor stonecrop), 170
Sedum 'Vera Jamison' (stonecrop), *8*
seedheads, 21, 137
seedpods, 21
'Senetti' (*Pericallis* 'Senetti'), *101*
sepal, 13, 216
See also names of specific plants
shapes, flower, 14
Shasta daisy (*Leucanthemum* x *superbum* 'Banana Cream'), 147
shieldleaf (*Astilboides tabularis*), 128
See also *Rodgersia astilboides* (shieldleaf)
shooting stars (*Dodecatheon pulchellum* ssp. *alaskanum*), *37*, 131, *131*
shoulder-season plants, 37
showy geranium (*Geranium* x *magnificum*), 137
shrub clematis (*Clematis integrifolia*), 130
shrubs, 179–193
deciduous, 179
evergreen, 179
shopping tips, 87–88
winter-display, 41
See also names of specific types of shrubs
Siberian iris (*Iris sibirica*), *26*, 142, 145–46, *146*

motion, 29
Siberian squill (*Scilla siberica*), 113
silver-edged primrose (*Primula marginata*), 159, *159*
size
  flower, 13
  foliage, 9
  plant, 35, 49
slug, 81
snake's head (*Fritillaria meleagris*), 109
snow-in-summer (*Cerastium tomentosum*), 14
snowmound spirea (*Spiraea nipponica* 'Snowmound'), 191–92
soil
  clay, 67
  fertility, 65, 67
  loam, 67
  moist, 69
  moist, well-drained, 69
  pH, 65, 67
  physical makeup, 62, 67
  sand, 67
  silt, 67
  testing, 63, 67
  type, 65, 67
solitary clematis (*Clematis integrifolia*), 130
Sorbaria sorbifolia, 188
Sorbaria sorbifolia 'Sem' (ash leaf spirea Sem; false spirea Sem), *188,* 188–89, *189*
sound (music), garden, 1–2
species, 97, 98, 216
  See also names of specific plants
spider mites, 80
spiked speedwell (*Veronica spicata* 'Royal Candles'), 175–76
spiky delphinium, *24*
spiky golden foxtail grass, *116*
spiky speedwell (*Veronica spicata* 'Red Fox'), *15*
Spiraea (spirea), 189–92
  See also names of plants in the Spirea/spirea family
spirea (*Spiraea*), 189–92
Spiraea fritschiana 'Pink Parasols' (pink parasols spirea), *190,* 190–91, *191*
Spiraea fritschiana 'Wilma' (pink parasols spirea), *190,* 190–91
  See also *Spiraea fritschiana* 'Pink Parasols' (pink parasols spirea)
Spiraea nipponica 'Snowmound' (snowmound spirea), *190,* 191–92
Spiraea x *bumalda* 'Denistar' (First Editions Superstar Spirea), *189,* 189–90, *190*
sport, 216
  See also names of specific plants
spotted deadnettle (*Lamium maculatum* 'White Nancy'), *10*
spring squill (*Scilla*), 112–13
Stachys byzantina 'Helen von Stein' (lambs' ears), *6,* 9
  bees and, 14
staking, plant, 57

stamen, 13, *212*, 216
  See also names of specific plants
standard dwarf bearded iris, 144–45
stems, 17
  herbaceous perennials, 19
stonecrop (*Sedum*), 167–70
  See also names of plants in the stonecrop/Sedum family
stonecrop (*Sedum kamtschaticum*), *47*
stonecrop (*Sedum* 'Vera Jamison'), *8*
straw flowers (*Bracteantha bracteata* 'Mohave Deep Rose'), *101*
subspecies, 98
sucker, 51, 216
  See also names of specific plants
suitability, plant, 35, 55, 57
sun cascade yellow flag iris (*Iris pseudacorus* 'Sun Cascade'), 143–44
Syringa pubescens subsp. *patula* 'Miss Kim' ('Miss Kim' lilac), *40*
Syringa vulgaris 'Sensation' (lilac), *30*

Tanacetum coccineum 'Robinson's Hybrid' (painted daisy), 171, *171*
Tanacetum coccineum 'Robinson's Pink,' *171*
Tatarian dogwood (*Cornus alba* 'Gouchaultii'), 183–84
tender perennial, 216
  See also names of specific plants
texture
  flower, 14
  foliage, 9
threadleaf coreopsis (*Coreopsis verticillata* 'Zagreb'), *34*
thyme (*Thymus*), 171–73
Thymus (thyme), 171–73
  See also names of plants in the Thymus/thyme family
Thymus coccineus (red creeping thyme), 173
Thymus praecox 'Highland Cream' (creeping thyme), 172, *172*
Thymus pseudolanuginosus (woolly thyme), *172,* 172–73
Thymus serpyllum 'Coccineus' (red creeping thyme), 173, *173*
Thymus serpyllum 'Pink Chintz' (creeping thyme), 173, *173*
Tibetan primrose (*Primula florindae*), 158–59
tickseed (*Bidens ferulifolia*), 103
  See also bidens (*Bidens ferulifolia*)
Toba hawthorn (*Crataegus* x *mordenensis* 'Toba'), 199–200
trees, 195–207
  architecture, 195
  bark, 195
  evergreens, 195
  shopping tips, 88–89, 91
  wildlife danger to, 195
  winter-display, 41
  See also names of specific trees
trilliums, 107

Trollius (globeflower; trollius), 173–75
See also names of plants in the Trollius/globeflower family
trollius (*Trollius*), 173–75
Trollius chinensis 'Golden Queen' (globeflower; queen of the buttercups), 174, *174*
Trollius europaeus, 173, 174, *174*
Trollius pumilus (dwarf globeflower; dwarf trollies), 174, *174*
Trollius x *cultorum* 'Alabaster,' 175
Trollius x *cultorum* 'New Moon' (New Moon globeflower), 175, *175*
Tubergeniana squill, 112
tubers, 107
Tulipa humilis 'Persian Pearl,' *106*
Tulipa 'Princess Irene,' *111*
tulips, *85*, 107
burrowing animals and, 83–84
turgid, 80, 216
Tweedy's Lewisia (*Lewisia tweedyi*), 148–49

U.S. Department of Agriculture (USDA)
hardiness zone map, 66
umbels, 125, 216

Valerie Finnis grape hyacinth (*Muscari neglectum* 'Valerie Finnis'), 110–11
variegated feather reed grass (*Calamagrostis* x *acutiflora* 'Avalanche'; *Calamagrostis* x *acutiflora* 'Overdam'), 118–19, *133*
variegated red twig dogwood (*Cornus alba* 'Bailhalo'; *Cornus alba* 'Elegantissima'), 183
variety, 97, 98, 216
See also names of specific vines
Verbascum (mullein), *18*
verbena, 83
veronica, 14, *24*
bees and, 14
Veronica spicata 'Red Fox' (spiky speedwell), *15*
Veronica spicata 'Royal Candles' (spiked speedwell), *175,* 175–76
Veronicastrum sibiricum, 133
Veronicastrum virginicum 'Apollo' (Culver's root), 176, *176*
Viburnum trilobum 'Bailey Compact' (Bailey compact cranberry; compact American highbush cranberry), 192–93, *193*
vigor, plant, 35, 55
vines
annual, 209
herbaceous, 209
perennial, 209–13
shopping tips, 91
woody, 209
See also names of specific vines
Viola 'Etain,' *32, 122*
Viola labradorica (woodland viola), 159

violas, *87*
wasps, parasitic, 80
weeds, 81
weeping white birch (*Betula pendula*), motion of, 29
weeping willows, motion of, 29
western mugwort (*Artemisia ludoviciana* 'Valerie Finnis'), 126–27
white birch (*Betula papyrifera*), 17, 197–98
See also paper birch (*Betula papyrifera*)
white flies, 80
white sage (*Artemisia ludoviciana* 'Valerie Finnis'), 126–27
white shooting star (*Dodecatheon dentatum*), 131, *131*
wild hyacinth (*Camassia quamash*), 177
wild iris (*Iris setosa*), 145
wildlife
attraction to plants, 57, 83–84, 182, 188, 189, 195
wind conditions, 63
winter display plants, 35, 41
shrubs, 41
trees, 41
witch hazel (*Hamamelis*), 188
wolfsbane (*Aconitum* x *cammarum* 'Bicolor'), 124–25
See also bicolor monkshood (*Aconitum* x *cammarum* 'Bicolor')
wood sage (*Salvia nemorosa* 'May Night'), 165
woodland viola (*Viola labradorica*), 159
woolly thyme (*Thymus pseudolanuginosus*), 172–73

yarrow, *87*
yarrow (*Achillea millefolium* 'Paprika'), *34*
yarrow (*Achillea millefolium* 'Terracotta'), *42*
yellow flag iris (*Iris pseudacorus*), 142
yellow-twigged dogwood (*Cornus sericea* 'Flaviramea'), 41

Zigadenus elegans (camas wand lily; death camas), 177, *177*